ELECTRONIC COMMUNICATION

AUTHOR

Dr. Vimal Bhatia

Professor,

Department of Electrical Engineering,

IIT Indore, Madya Pradesh

REVIEWER

Dr. Manav Bhatnagar

Professor,

Department of Electrical Engineering,

IIT Delhi

KHANNA BOOK PUBLISHING CO. (P) Ltd.

4C/4344, Ansari Road, Darya Ganj, New Delhi, India

All India Council for Technical Education

Nelson Mandela Marg, Vasant Kunj, New Delhi, India

Dear Readers,

To prevent the piracy, this book is secured with HIGH SECURITY HOLOGRAM on the front title cover. In case you don't find the hologram on the front cover title, please write us to at contact@khannabooks.com or whatsapp us at +91-99109 09320 and avail special gift voucher for yourself.

Specimen of Hologram on front Cover title:

Moreover, there is a SPECIAL DISCOUNT COUPON for you with EVERY HOLOGRAM.

How to avail this SPECIAL DISCOUNT:

Step 1: Scratch the hologram

Step 2: Under the scratch area, your "coupon code" is available

Step 3: Logon to www.khannabooks.com

Step 4: Use your "coupon code" in the shopping cart and get your copy at a special discount

Step 5: Enjoy your reading!

ISBN: 978-93-5538-465-2

Book Code: UG422EN

Electronic Communication

by Dr. Vimal Bhatia

First Edition: 2026

Originally Published by:

All India Council for Technical Education, Ministry of Education, GOI

Exclusively Published & Distributed by:

Khanna Book Publishing Co. (P) Ltd.
4C/4344, Ansari Road, Daryaganj,
New Delhi (INDIA)

Visit us at: **www.khannabooks.com**
Write us at: **contact@khannabooks.com**

To view complete list of books, Please scan the QR Code: KPH

Printed in Bharat.

प्रो. टी. जी. सीताराम
अध्यक्ष
Prof. T. G. Sitharam
Chairman

सत्यमेव जयते

आज़ादी का अमृत महोत्सव

अखिल भारतीय तकनीकी शिक्षा परिषद्
(भारत सरकार का एक सांविधिक निकाय)
(शिक्षा मंत्रालय, भारत सरकार)
नेल्सन मंडेला मार्ग, वसंत कुंज, नई दिल्ली-110070
दूरभाष : 011-26131498
ई-मेल : chairman@aicte-india.org

ALL INDIA COUNCIL FOR TECHNICAL EDUCATION
(A STATUTORY BODY OF THE GOVT. OF INDIA)
(Ministry of Education, Govt. of India)
Nelson Mandela Marg, Vasant Kunj, New Delhi-110070
Phone : 011-26131498
E-mail : chairman@aicte-india.org

FOREWORD

Engineers are the backbone of any modern society. They are the ones responsible for the marvels as well as the improved quality of life across the world. Engineers have driven humanity towards greater heights in a more evolved and unprecedented manner.

The All India Council for Technical Education (AICTE), have spared no efforts towards the strengthening of the technical education in the country. AICTE is always committed towards promoting quality Technical Education to make India a modern developed nation emphasizing on the overall welfare of mankind.

An array of initiatives has been taken by AICTE in last decade which have been accelerated now by the National Education Policy (NEP) 2020. The implementation of NEP under the visionary leadership of Hon'ble Prime Minister of India envisages the provision for education in regional languages to all, thereby ensuring that every graduate becomes competent enough and is in a position to contribute towards the national growth and development through innovation & entrepreneurship.

One of the spheres where AICTE had been relentlessly working since past couple of years is providing high quality original technical contents at Under Graduate & Diploma level prepared and translated by eminent educators in various Indian languages to its aspirants. For students pursuing 2nd year of their Engineering education, AICTE has identified 87 books, which shall be translated into 12 Indian languages - Hindi, Tamil, Gujarati, Odia, Bengali, Kannada, Urdu, Punjabi, Telugu, Marathi, Assamese & Malayalam. In addition to the English medium, books in different Indian Languages are going to support the students to understand the concepts in their respective mother tongue.

On behalf of AICTE, I express sincere gratitude to all distinguished authors, reviewers and translators from the renowned institutions of high repute for their admirable contribution in a record span of time.

AICTE is confident that these outcomes based original contents shall help aspirants to master the subject with comprehension and greater ease.

T. G. Sitharam

(Prof. T. G. Sitharam)

ACKNOWLEDGEMENT

The authors are grateful to the authorities of AICTE, particularly Prof. T. G. Sitharam, Chairman; Prof. M. P. Poonia, Vice-Chairman; Prof. Rajive Kumar, Member-Secretary and Dr Amit Kumar Srivastava, Director, Faculty Development Cell for their planning to publish the books on Electronic Communication. We sincerely acknowledge the valuable contributions of the reviewer of the book Dr. Manav Bhatnagar, Professor, IIT Delhi for making it students' friendly and giving a better shape in an artistic manner.

This book is an outcome of various suggestions of AICTE members, experts and authors who shared their opinion and thought to further develop the engineering education in our country. Acknowledgements are due to the contributors and different workers in this field whose published books, review articles, papers, photographs, footnotes, references and other valuable information enriched us at the time of writing the book.

Dr. Vimal Bhatia

PREFACE

The book titled "Electronic Communication" is an outcome of the rich experience of our teaching of basic analog and digital communication courses. The initiation of writing this book is to expose basic knowledge of communication related concepts to the engineering students, the fundamentals as well as enable them to get an insight of the subject. Keeping in mind the purpose of wide coverage as well as to provide essential supplementary information, we have included the topics recommended by AICTE, in a very systematic and orderly manner throughout the book. Efforts have been made to explain the fundamental concepts of the subject in the simplest possible way.

During the process of preparation of the manuscript, we have considered the various standard text books and accordingly we have developed sections like critical questions, solved and supplementary problems etc. While preparing the different sections emphasis has also been laid on definitions and theorems and also on comprehensive synopsis of formulae for a quick revision of the basic principles. The book covers all types of medium and advanced level problems and these have been presented in a very logical and systematic manner. The gradations of those problems have been tested over many years of teaching to a wide variety of students.

Apart from illustrations and examples as required, we have enriched the book with numerous solved problems in every unit for proper understanding of the related topics. It is important to note that in all the books, we have included the relevant laboratory practical experiments to ensure that our readers get a hands on training of the topics discussed. This is a salient feature of this book.

In addition, besides some essential information for the users under the heading "Know More" we have clarified some essential basic information in the appendix and annexure section.

As far as the present book is concerned, "Electronic Communication" is meant to provide a thorough grounding in communication engineering on the topics covered. This part book will prepare engineering students to apply the knowledge of communication engineering to address challenges of the future and help them research on more advanced topics such as 5G and beyond 5G, MIMO, beamforming etc. The subject matters are presented in a constructive manner so that an Engineering degree prepares students to work in different sectors or in national laboratories at the very forefront of technology. -

We sincerely hope that the book will inspire the students to learn and discuss the ideas behind basic principles of engineering physics and will surely contribute to the development of a solid foundation of the subject. We would be thankful to all beneficial comments and suggestions which will contribute to the improvement of the future editions of the book. It gives us immense pleasure to place this book in the hands of the teachers and students. It was indeed a big pleasure to work on different aspects covering in the book. To place this book in the hands of the teachers and students. It was indeed a big pleasure to work on different aspects covering in the book.

Dr. Vimal Bhatia

OUTCOME BASED EDUCATION

For the implementation of an outcome based education the first requirement is to develop an outcome based curriculum and incorporate an outcome based assessment in the education system. By going through outcome based assessments evaluators will be able to evaluate whether the students have achieved the outlined standard, specific and measurable outcomes. With the proper incorporation of outcome based education there will be a definite commitment to achieve a minimum standard for all learners without giving up at any level. At the end of the programme running with the aid of outcome based education, a student will be able to arrive at the following outcomes:

PO1. **Engineering knowledge:** Apply the knowledge of mathematics, science, engineering fundamentals, and an engineering specialization to the solution of complex engineering problems.

PO2. **Problem analysis:** Identify, formulate, review research literature, and analyze complex engineering problems reaching substantiated conclusions using first principles of mathematics, natural sciences, and engineering sciences.

PO3. **Design / development of solutions:** Design solutions for complex engineering problems and design system components or processes that meet the specified needs with appropriate consideration for the public health and safety, and the cultural, societal, and environmental considerations.

PO4. **Conduct investigations of complex problems:** Use research-based knowledge and research methods including design of experiments, analysis and interpretation of data, and synthesis of the information to provide valid conclusions.

PO5. **Modern tool usage:** Create, select, and apply appropriate techniques, resources, and modern engineering and IT tools including prediction and modeling to complex engineering activities with an understanding of the limitations.

PO6. **The engineer and society:** Apply reasoning informed by the contextual knowledge to assess societal, health, safety, legal and cultural issues and the consequent responsibilities relevant to the professional engineering practice.

PO7. **Environment and sustainability:** Understand the impact of the professional engineering solutions in societal and environmental contexts, and demonstrate the knowledge of, and need for sustainable development.

PO8. **Ethics:** Apply ethical principles and commit to professional ethics and responsibilities and norms of the engineering practice.

PO9. **Individual and team work:** Function effectively as an individual, and as a member or leader in diverse teams, and in multidisciplinary settings.

PO10. Communication: Communicate effectively on complex engineering activities with the engineering community and with society at large, such as, being able to comprehend and write effective reports and design documentation, make effective presentations, and give and receive clear instructions.

PO11. Project management and finance: Demonstrate knowledge and understanding of the engineering and management principles and apply these to one's own work, as a member and leader in a team, to manage projects and in multidisciplinary environments.

PO12. Life-long learning: Recognize the need for, and have the preparation and ability to engage in independent and life-long learning in the broadest context of technological change.

COURSE OUTCOMES

After completion of the course the students will be able to:

CO-1: Describe communication and design of transmitter and receiver.

CO-2: Explain different modulation techniques in analog and digital communication.

CO-3: Apply probability for the analysis bit error rate un communication

CO-4: Apply signal detection techniques at the receiver for reconstructing the signal.

CO-5: Apply different digital mdulation techniques such as PSK, FSK.

CO-6: Learn the electronic communication and its application in real life.

Course Outcomes	Expected Mapping with Programme Outcomes *(1- Weak Correlation; 2- Medium correlation; 3- Strong Correlation)*											
	PO-1	PO-2	PO-3	PO-4	PO-5	PO-6	PO-7	PO-8	PO-9	PO-10	PO-11	PO-12
CO-1	3	2	2	2	1	-	-	-	-	-	-	-
CO-2	3	1	2	1	-	-	-	-	-	-	-	-
CO-3	3	3	2	1	-	-	-	-	-	-	-	-
CO-4	3	3	3	2	1	-	-	-	-	-	-	-
CO-5	3	3	2	1	1	-	-	-	-	-	-	-
CO-6	3	3	3	2	1	-	-	-	-	-	-	-

GUIDELINES FOR TEACHERS

To implement Outcome Based Education (OBE) knowledge level and skill set of the students should be enhanced. Teachers should take a major responsibility for the proper implementation of OBE. Some of the responsibilities (not limited to) for the teachers in OBE system may be as follows:

- Within reasonable constraint, they should manoeuvre time to the best advantage of all students.
- They should assess the students only upon certain defined criterion without considering any other potential ineligibility to discriminate them.
- They should try to grow the learning abilities of the students to a certain level before they leave the institute.
- They should try to ensure that all the students are equipped with the quality knowledge as well as competence after they finish their education.
- They should always encourage the students to develop their ultimate performance capabilities.
- They should facilitate and encourage group work and team work to consolidate newer approach.
- They should follow Blooms taxonomy in every part of the assessment.

Bloom's Taxonomy

Level	Teacher should Check	Student should be able to	Possible Mode of Assessment
Create	Students ability to create	Design or Create	Mini project
Evaluate	Students ability to justify	Argue or Defend	Assignment
Analyse	Students ability to distinguish	Differentiate or Distinguish	Project/Lab Methodology
Apply	Students ability to use information	Operate or Demonstrate	Technical Presentation/ Demonstration
Understand	Students ability to explain the ideas	Explain or Classify	Presentation/Seminar
Remember	Students ability to recall (or remember)	Define or Recall	Quiz

GUIDELINES FOR STUDENTS

Students should take equal responsibility for implementing the OBE. Some of the responsibilities (not limited to) for the students in OBE system are as follows:

- Students should be well aware of each UO before the start of a unit in each and every course.
- Students should be well aware of each CO before the start of the course.
- Students should be well aware of each PO before the start of the programme.
- Students should think critically and reasonably with proper reflection and action.
- Learning of the students should be connected and integrated with practical and real life consequences.
- Students should be well aware of their competency at every level of OBE.

LIST OF ABBREVIATIONS

General Terms			
Abbreviations	**Full form**	**Abbreviations**	**Full form**
AWGN	Additive white Gaussian noise	MMSE	Minimum mean square error
BCJR	Bahl-Cocke-Jelinek-Raviv	MSK	Minimum shift keying
BPF	Band pass filter	MUX	Multiplexer
BPSK	Binary phase shift keying	NBFM	Narrow band frequency modulation
BW	Bandwidth	NBPM	Narrow band phase modulation
CDF	Cumulative distribution function	NRZI	Non-return to zero inverted
CPFSK	Continuous phase frequency shift-keying	PAM	Pulse amplitude modulation
CPM	Continuous phase modulation	PCM	Pulse code modulation
DEMUX	Demultiplexer	PDF	Probability density function
DFE	Decision feedback equalizer	PLL	Phase locked loop
DM	Delta modulation	PPM	Pulse position modulation
DPCM	Differential pulse code modulation	PSD	Power spectral density
DSB	Double side band	PSK	Phase shift keying
DSBSC	Double side band suppressed carrier	PWM	Pulse width modulation
FDM	Frequency division multiplexing	QAM	Quadrature amplitude modulation
FM	Frequency modulation	QPSK	Quadrature phase shift keying
FSK	Frequency shift keying	SNR	Signal-to-noise ratio
HPF	High pass filter	SSB	Single side band
ISI	Inter symbol interference	TDM	Time division multiplexing
LPF	Low pass filter	VCO	voltage control oscillator
LTI	Linear time invariant	VSB	Vestigial Side band
MAP	Maximum posteriori probability	WBFM	Wideband frequency modulation
ML	Maximum likelihood	WSS	Wide-sense stationary
MLSD	Maximum likelihood sequence detection	ZF	Zero forcing
MLSE	Maximum likelihood sequence estimation		

LIST OF FIGURES

Unit 1: Analog Modulation

TABLE OF CONTENTS

1 Analog Modulation Techniques

UNIT SPECIFICS

Through this unit we have discussed the following aspects:

- *Review of signals and systems,*
- *Frequency domain representation of signals,*
- *Principles of Amplitude Modulation Systems- DSB, SSB and VSB modulations.*
- *Angle Modulation,*
- *Representation of FM and PM signals,*
- *Spectral characteristics of angle modulated signals.*

The practical applications of the topics are discussed for generating further curiosity and creativity as well as improving problem solving capacity.

Besides giving a large number of multiple-choice questions as well as questions of short and long answer types marked in two categories following lower and higher order of Bloom's taxonomy, assignments through a number of numerical problems, a list of references and suggested readings are given in the unit so that one can go through them for practice. It is important to note that for getting more information on various topics of interest some QR codes have been provided in different sections which can be scanned for relevant supportive knowledge.

After the related practical, based on the content, there is a "Know More" section. This section has been carefully designed so that the supplementary information provided in this part becomes beneficial for the users of the book. This section mainly highlights the initial activity, examples of some interesting facts, analogy, history of the development of the subject focusing the salient observations and finding, timelines starting from the development of the concerned topics up to the recent time, applications of the subject matter for our day-to-day real life or/and industrial applications on variety of aspects, case study related to environmental, sustainability, social and ethical issues whichever applicable, and finally inquisitiveness and curiosity topics of the unit.

RATIONALE

This fundamental unit on analog modulation techniques helps students to get a primary idea about the different types of amplitude and angle modulation techniques and their application in communication process. We have started with basics and pre-requisites required for the topic. First, signals and systems is thoroughly revised before moving on to the more complex topics. Then we have covered the amplitude modulation, its types and demodulation process. Similarly, then angle modulation and demodulation is covered. In angle modulation phase and frequency modulation are thoroughly explained. The chapter is concluded with some MCQs and numerical questions.

Modulation is the most important step in transmitting an information bearing signal over a bandpass channel, such as a telephone line, satellite channel etc.. It accomplishes the task of shifting the range of frequencies of the signal to some another frequency range suitable for transmission. We will learn more about the modulation process in this chapter.

PRE-REQUISITES

Mathematics: Calculus (Class XII)

Physics: Basic of Communication (Class XII)

UNIT OUTCOMES

List of outcomes of this unit is as follows:

U1-O1: Describe basics of signals and systems and frequency representation of signals.

U1-O2: Explain the amplitude modulation systems

U1-O3: Describe the angle modulation and their signal representation

U1-O4: The spectral characteristics of angle modulated signal is explained.

U1-O5: Application of modulation for real life.

This table below need to be filled after discussion as marked with red.

Unit-1 Outcomes	EXPECTED MAPPING WITH COURSE OUTCOMES (1- Weak Correlation; 2- Medium correlation; 3- Strong Correlation)					
	CO-1	CO-2	CO-3	CO-4	CO-5	CO-6
U1-O1	1	2	3	-	3	1
U1-O2	1	1	2	2	1	-
U1-O3	2	1	3	1	2	1
U1-O4	-	-	3	1	2	2
U1-O5	3	3	3	-	3	1

1.1 INTRODUCTION

In electronics communication, the signal is defined as the quantity representing information such as electric current or electromagnetic waves. For example, the human voice is a form of the speech signal, the speed of a car.

The system represents the mathematical relationship between the input and output signals.

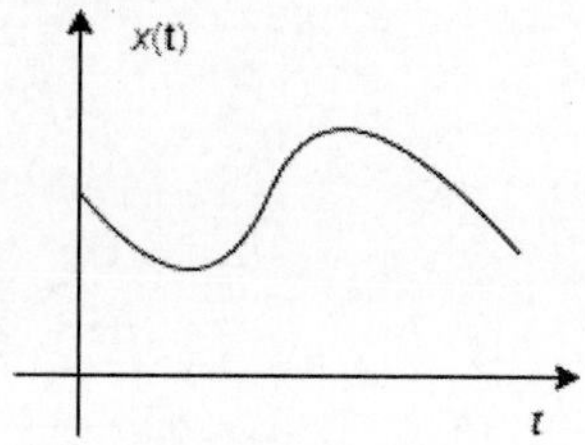

Fig. 1.1 Message signal

1.2 ELEMENTARY FUNCTIONS

In this section, we will learn about some elementary functions and their mathematical expressions in brief. In any real-world engineering applications, after expressing the physical system into a suitable mathematical mode we first test our designed system with a certain set of basic pre-defined inputs that imitate the characteristics of the input we want to apply to the system. The response of the system is then recorded and analyzed by using various mathematical tools such as Fourier transforms.

But it is not practically possible to model every random input mathematically. Thus, we take help of certain elementary signals which replicate most inputs we want to apply to the system. For example, if the input is a sudden blow or have high impact in a very short duration of time, we can model this using an impulse signal; similarly, if it is a input with some finite magnitude that lasts for a very long duration or for an infinite duration we can model it as a step signal. Like this, we will study following signals which found extensive use in our applications.

1.2.1 Unit Step Function

The unit step function u(t) shown in Fig. (1.2) is defined as

$$u(t) = \begin{cases} 1, & t \geq 0 \\ 0, & t < 0 \end{cases}$$

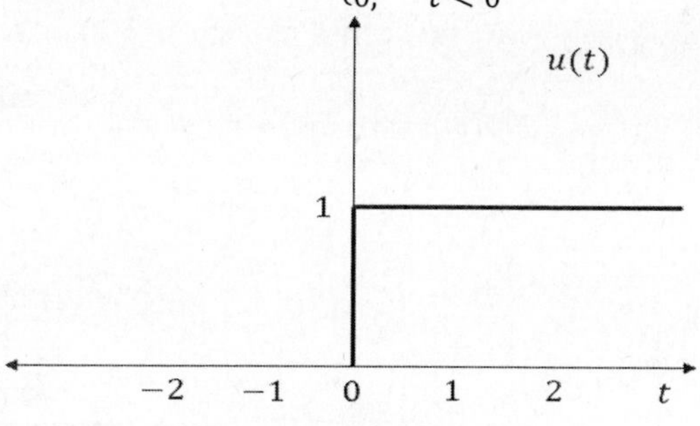

Fig. 1.2 Unit step function

1.2.2 Dirac-Delta Function

The dirac-delta function $\delta(t)$ as shown in Fig. (1.3) is defined as

$$\int_{-\infty}^{\infty} \phi(t)\delta(t)\, dt = \phi(0) \tag{1.1}$$

where $\phi(t)$ is any testing function varying with time and continuous at $t = 0$.

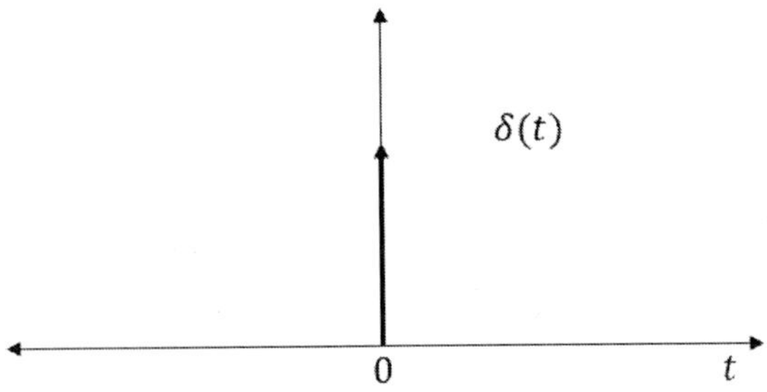

Fig. 1.3 Dirac-delta function

If the delta function is delayed by t_0 Eq. (1.1) can be written as

$$\int_{-\infty}^{\infty} \phi(t)\delta(t-t_0)\,dt = \phi(t_0) \tag{1.2}$$

where $\phi(t)$ is any function continuous at $t = t_0$.

Some properties of Dirac-delta function are as follows

1. $\delta(t-t_0) = 0$, if $t \neq t_0$.
2. $\phi(t)\delta(t-t_0) = \phi(t_0)\delta(t-t_0)$, where $\phi(t)$ is continuous at $t = t_0$.
3. $\phi(t)\delta(t) = \phi(0)\delta(t)$, where $\phi(t)$ is continuous at $t = 0$.
4. $\delta(bt) = \frac{1}{|b|}\delta(t), b \neq 0$.
5. $\delta(-t) = x(t)$
6. $x(t) * \delta(t) = x(t)$
7. $\int_{t_1}^{t_2} \delta(t-t_0)\,dt = 1$, where $t_1 < t_0 < t_2$.
8. Spectrum of $\delta(t)$ extends uniformly over the entire frequency range as shown in Fig. (1.5). We can also simply derive the Fourier transform of $\delta(t)$ as well

$$\mathcal{F}[\delta(t)] = \int_{-\infty}^{\infty} \delta(t)e^{-j\omega t}\,dt = 1 \tag{1.3}$$

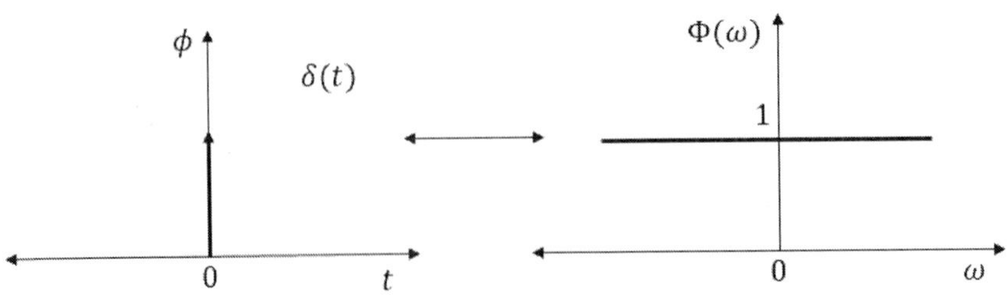

Fig. 1.4 Dirac-delta function and its spectrum

Similarly, spectrum of a constant signal is a delta function as shown in Fig. (1.5).

$$1 \leftrightarrow 2\pi\delta(\omega)$$

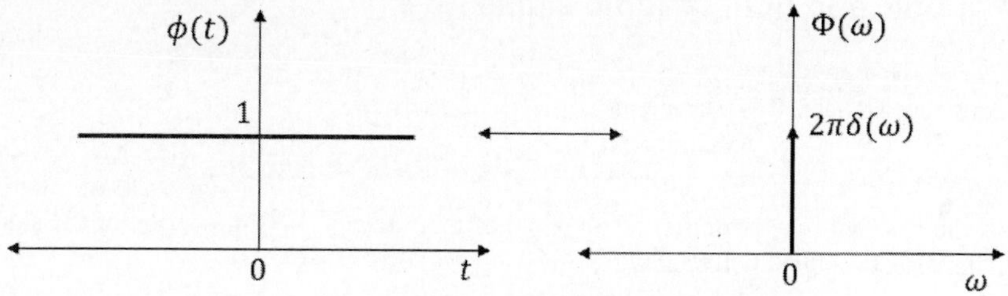

Fig. 1.5 Constant signal and its spectrum

1.2.3 Unit Ramp Function

The unit ramp function $r(t)$ as shown in Fig. (1.6) is defined as

$$r(t) = \begin{cases} t, & t \geq 0 \\ 0, & t < 0 \end{cases}$$

Unit ramp signal can also be obtained from the unit step signal by integrating it.

$$r(t) = \int_{-\infty}^{t} u(\tau)\, d\tau$$

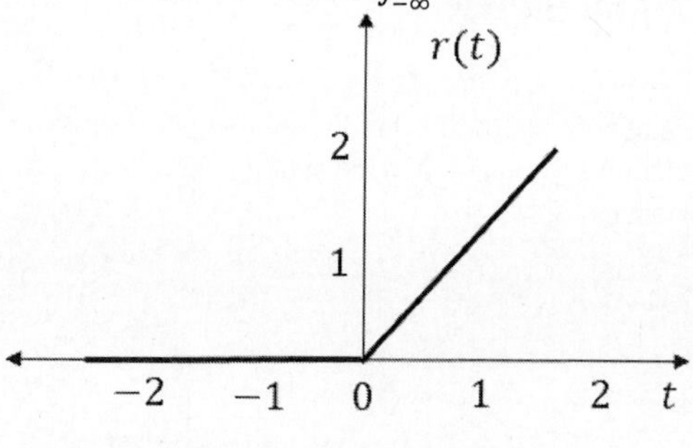

Fig. 1.6 Unit ramp function

1.3 SIGNAL CLASSIFICATION

1.3.1 Continuous-time and Discrete-time Signals

These signals are classified based on independent variables. The signal defined for every instant of time (t) is called a continuous signal, and signal defined for a discrete instant of independent variable [n] is known as a discrete-time signal. e.g. speech signal is a continuous-time signal and marks of a student in a class is a discrete-time signal.

1.3.2 Periodic and non-periodic signal

A continuous-time signal x(t) is periodic if

$$x(t) = x(t + mT), \text{ for all 't'}$$

where T is the constant time period or also known as the fundamental time period of the signal, and m is an integer. Otherwise, the signal will aperiodic.

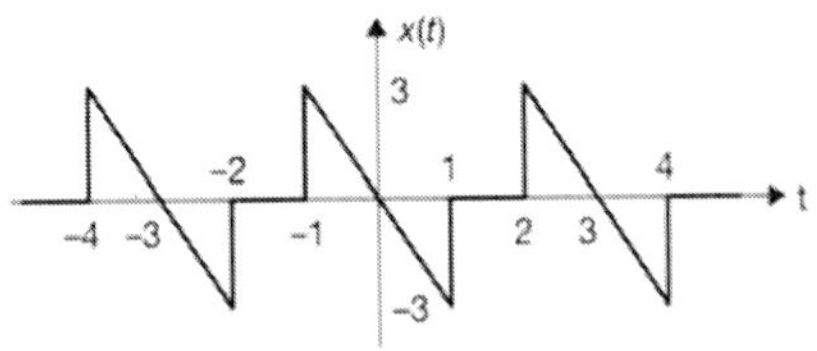

Fig. 1.7 periodic signal.

1.3.3 Analog and Digital Signal

This classification of the signal is based on the amplitude of the signal. If the amplitude of the signal takes all possible real values in the continuous range, then the signal is known as analog signal. While the digital signal is one which takes specific values.

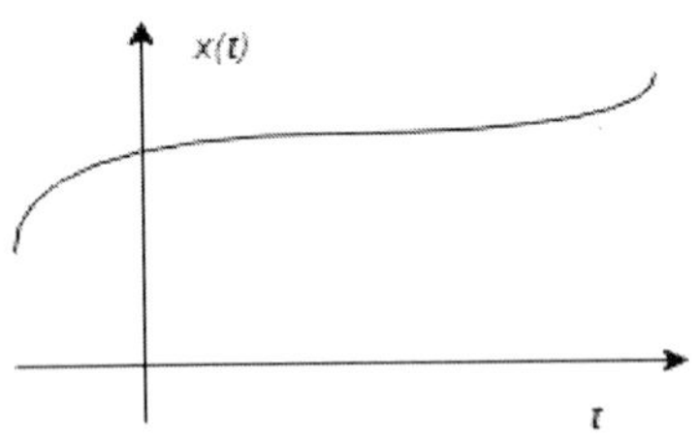

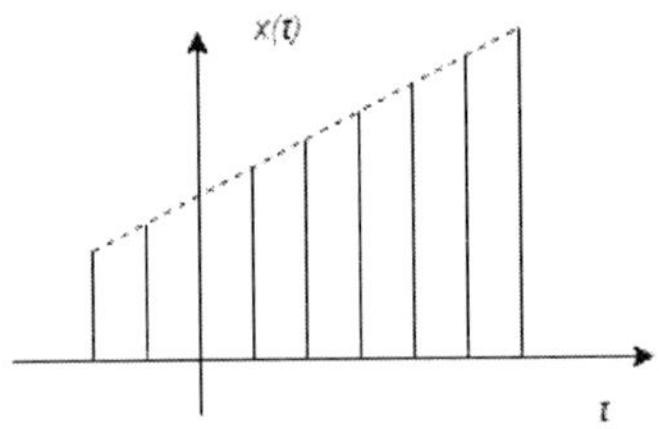

Fig. 1.8 Continuous analog signal

Fig. 1.9 Discrete signal

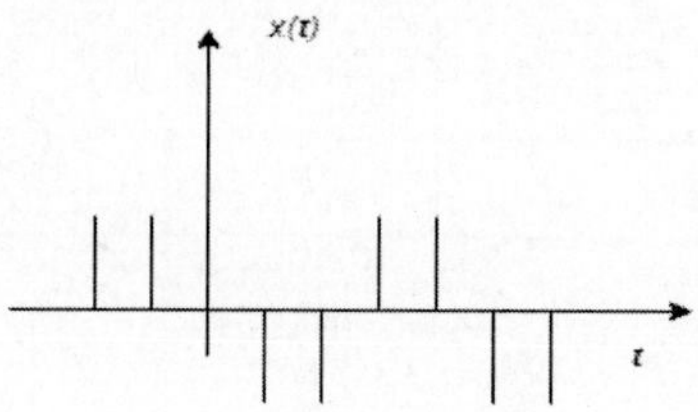

Fig. 1.10 Digital signal

1.3.4 Energy and Power Signal

The size of the signal can be described in one of these terms: energy and power signal. The energy of the signal $x(t)$ can be expressed as

$$E_x = \int_{-\infty}^{\infty} |x(t)|^2 dt \tag{1.4}$$

If the integral given in Eq. (1.4) does not converge then the signal has infinite energy. A signal with finite amplitude and time-limitation ($|x(t)| < \infty$) has finite energy. Moreover, real communication signals contain finite amounts of energy since they are time-limited and have finite amplitudes. As signals such as sinusoids and complex exponential are periodic signals therefore the energy of such signals is infinite.

However, periodic signals have finite power. As power signals are those signals which have infinite energy. The general expression for power signal is

$$P_x = \lim_{T \to \infty} \frac{1}{2T} \int_{-T}^{T} |x(t)|^2 dt. \tag{1.5}$$

1.4 OPERATION OF SIGNALS

1. Time shifting
2. Time reversal
3. Time scaling

The time-scaled form of a signal, $x(t)$, is obtained by changing 't' to 'at', where 'a' is the scaling factor.

$$y(t) = x(at)$$

- $a > 1$ scaling is termed as compression of $x(t)$,
- $0 < a < 1$ is termed as expansion of $x(t)$.

Example 1.1:- Time Scaling of a rectangular pulse by scaling factor 2.

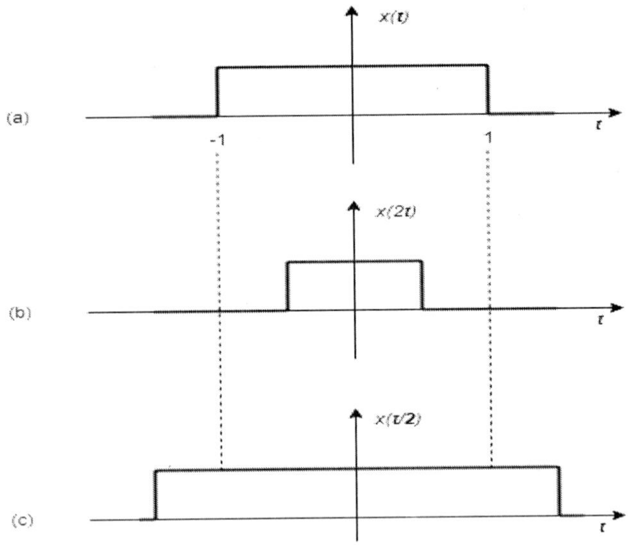

The time-shifted signal results in time delay or time advancement.

- Time delay signal $y(t) = x(t - T)$
- Time advancement signal $y(t) = x(t + T)$

Example 1.2: - Signal is time shifted by 'T'

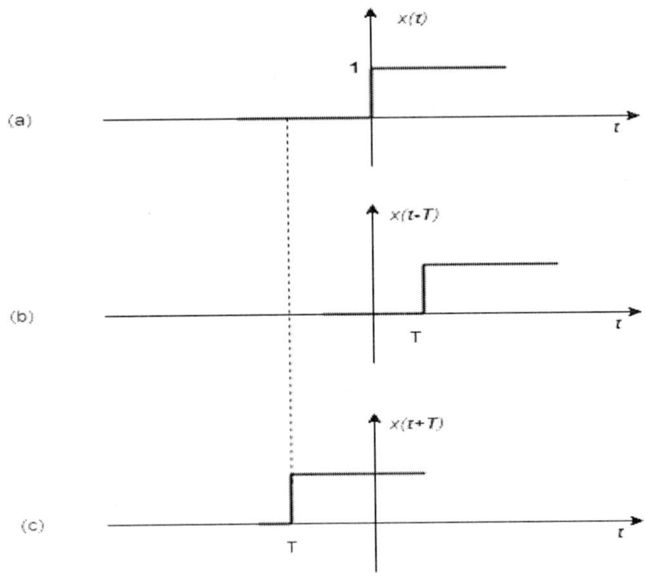

Time reversal of signal $x(t)$ is achieved by replacing 't' by '-t'. This is also known as folding or reflection about vertical axis. The common example of time reversal is mirror.

Example 1.3: - Time reversed signal.

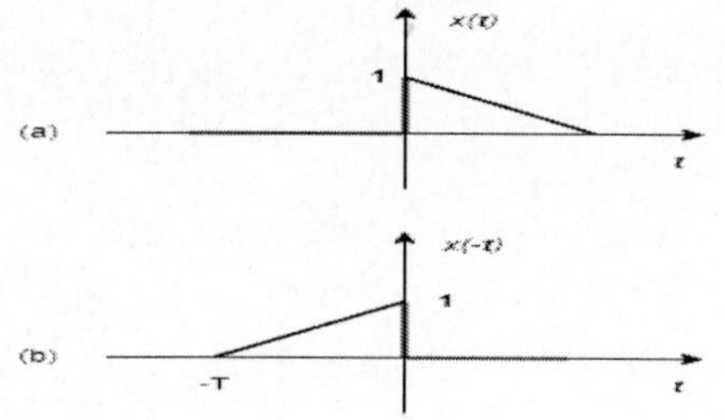

1.5 SYSTEMS

A system is defined as the mathematical relation between input and output signal.

There are three types of systems, Linear, Time-invariant and Time-variant systems.

1. A Linear system is based on the superposition principle. According to this principle, a linear combination of input signals produces a linear combination of equivalent output signals as its output. Let $x_1(t)$ and $x_2(t)$ are two input signals corresponds to output signals $y_1(t)$ and $y_2(t)$, respectively. Then, for input $a_1x_1(t) + a_2x_2(t)$ the output will be $a_1y_1(t) + a_2y_2(t)$ for linear system.
2. A time-invariant system is a system which does not change with time. Let y(t) is the output for input x(t), then for input *x(t-T)* the output will be *y(t-T)* for delay T. Otherwise the system is time-variant system.

1.5.1 Linear Time-Invariant Systems

Linear time-invariant (LTI) systems are system which depict properties of linear as well as time-invariant system. Also, LTI system is described by its impulse response. The impulse response is the output when the input is an impulse, i.e., a Dirac delta function.

If *h(t)* is the impulse response of an LTI system, then the output *x(t)* can be found as

$$y(t) = \int_{-\infty}^{\infty} x(\lambda)h(t-\lambda)d\lambda = x(t) * h(t) \tag{1.6}$$

where '*' is the convolution operation and it is defined in section (1.4.2).

1.5.2 Correlation and Convolution

In the previous section, we have learnt about the LTI systems where the mathematical operation convolution is used to express the relation between input and output of the LTI system. Convolution relates the input, output and impulse response of an LTI system as

$$y(t) = x(t) * h(t)$$

where $y(t)$ is the output of LTI system, x(t) is the input to the LTI system, and $h(t)$ is the impulse response of the LTI system. For continuous signals convolution can be computed as

$$y(t) = \int_{-\infty}^{\infty} x(\tau)h(t-\tau)d\tau$$

Similarly, for discrete signals, convolution operation is performed as

$$y(n) = \sum_{k=-\infty}^{\infty} x(k)h(n-k)$$

For ex., For the discrete signals convolution output is shown in Fig. (1.11)

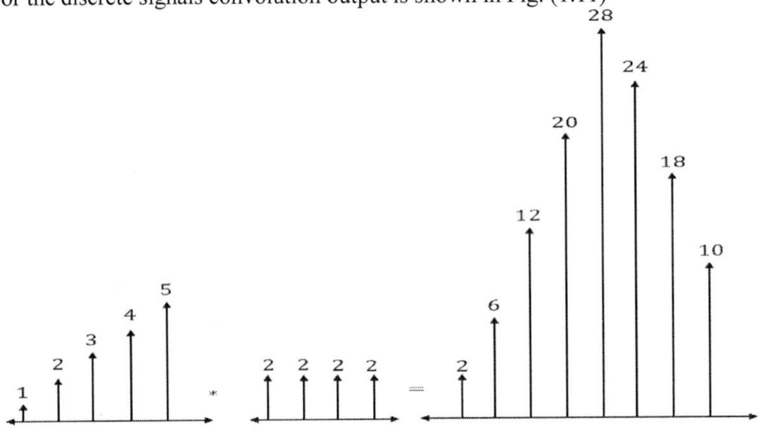

Fig. 1.11 Dirac-delta function and its spectrum

Correlation operation is also a convolution operation between two signals to measure the similarity between them. But there is a basic difference. Correlation of two signals is the convolution between one signal with the functional inverse version of the second signal

$$r(t) = \int_{-\infty}^{\infty} x(\tau)h(\tau-t)d\tau$$

$$r(t) = x(t) * h(-t)$$

Correlation is of two types

1. Auto correlation- It is defined as correlation of a signal with itself.

$$R_{xx}(\tau) = \int_{-\infty}^{\infty} x(\tau)x(\tau-t)d\tau$$

   Here τ is the delay parameter

2. Cross-correlation- It is defined as correlation of a signal with another signal. Or we can say that it is a measure of similarity for two different signals.

$$R_{xy}(\tau) = \int_{-\infty}^{\infty} x(\tau)y(\tau-t)d\tau$$

Properties of Convolution

In this section let's learn about the properties of convolution which would be useful throughout this course

1. Commutative Property
$$x(t) * h(t) = h(t) * x(t)$$
2. Distributive Property
$$h(t) * [x_1(t) + x_2(t)] = [h(t) * x_1(t)] + [h(t) * x_2(t)]$$
3. Associative Property
$$h(t) * [x_1(t) * x_2(t)] = [h(t) * x_1(t)] * x_2(t)$$
4. Shifting Property
$$x(t) * h(t) = y(t)$$
$$x(t) * h(t - t_0) = y(t - t_0)$$
$$x(t - t_0) * h(t) = y(t - t_0)$$
$$x(t - t_0) * h(t - t_1) = y(t - t_0 - t_1)$$
5. Convolution with delta
$$x(t) * \delta(t) = x(t)$$
6. Convolution of Unit step signal
$$u(t) * u(t) = r(t)$$
7. Scaling Property
If $x(t) * h(t) = y(t)$
then $x(at) * h(at) = \frac{1}{|a|} y(at)$
8. Differentiation of Output
if $y(t) = x(t) * h(t)$
then $\frac{dy(t)}{dt} = \frac{dx(t)}{dt} * h(t)$
or
$$\frac{dy(t)}{dt} = x(t) * \frac{dh(t)}{dt}$$

Example 1.4: - Compute response of continuous time LTI system with input x(t) = $e^{ax}\, u(t)$ and impulse response $h(t) = \delta(t)$.

Solution: - The response of LTI system is given as

$$y(t) = \int_{-\infty}^{\infty} x(\lambda) h(t - \lambda) d\lambda = x(t) * h(t)$$

given x(t) = $e^{at}\, u(t)$ and $h(t) = \delta(t)$

$$y(t) = \int_{-\infty}^{\infty} e^{a\lambda}\, u(\lambda)\delta(t - \lambda)\, d\lambda$$

$$y(t) = e^{at}\, u(t)$$

1.5.3 Hilbert Transform

Hilbert transform of a signal $x(t)$ defined as the transform in which phase angle of all the components of the signal is shifted by an angle of 90°, as shown in Fig. 1.12.

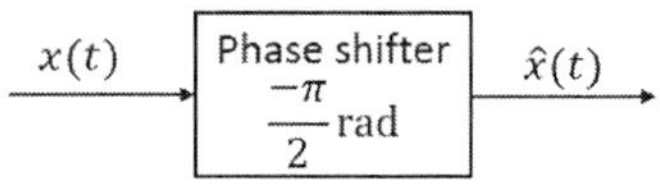

Fig. 1.12 Hilbert Transform

Mathematically, Hilbert transform of a signal $x(t)$ can be estimated as

$$\hat{x}(t) = \frac{1}{\pi}\int_{-\infty}^{\infty}\frac{x(k)}{t-k}dk \tag{1.7}$$

Similarly, the inverse Hilbert transform is given as

$$x(t) = \frac{-1}{\pi}\int_{-\infty}^{\infty}\frac{\hat{x}(k)}{t-k}dk \tag{1.8}$$

Properties of Hilbert transform:

1. $x(t)$ and $\hat{x}(t)$ have same amplitude spectrum.
2. $x(t)$ and $\hat{x}(t)$ same autocorrelation function.
$$E[x(t)x(t-\tau)] = E[\hat{x}(t)\hat{x}(t-\tau)]$$
3. The energy spectral density is same for both $x(t)$ and $\hat{x}(t)$.
$$\hat{X}(f) = -j\text{sgn}(f)X(f)$$
$$|X(f)| = |\hat{X}(f)|$$
4. $x(t)$ and $\hat{x}(t)$ are orthogonal.
$$\langle x(t)\hat{x}(t)\rangle = \int_{-\infty}^{\infty} x(t)\hat{x}^*(t)dt$$
$$= \int_{-\infty}^{\infty} X(f)\hat{X}^*(f)df$$
$$= \int_{-\infty}^{\infty} X(f)[-j\text{sgn}(f)\hat{X}(f)]^*df$$
$$= \int_{-\infty}^{\infty} j|X(f)|^2\text{sgn}(f)df$$
$$= 0$$
5. The Hilbert transform of $\hat{x}(t)$ is $-x(t)$ as can be seen from equation (1.7) and (1.8).
6. If Fourier transform exist, then Hilbert transform also exists for energy and power signals.

Example 1.5: Calculate the Hilbert transforms of the basic trigonometric functions. Suppose $\omega > 0$.

$$\mathcal{H}(\cos(\omega t)) = \mathcal{H}\left(\frac{e^{i\omega t}+e^{-i\omega t}}{2}\right) = \frac{\mathcal{H}(e^{i\omega t})+\mathcal{H}(e^{i(-\omega)t})}{2}$$

$$= \frac{-i\mathrm{sgn}(\omega)e^{i\omega t} - i\mathrm{sgn}(-\omega)e^{i(-\omega)t}}{2}$$

$$= \frac{-ie^{i\omega t} - ie^{-i\omega t}}{2} = \frac{e^{i\omega t}-e^{-i\omega t}}{2i} = \sin(\omega t)$$

$$\mathcal{H}(\sin(\omega t)) = \mathcal{H}\left(\cos\left(\omega t - \frac{\pi}{2}\right)\right) = \sin\left(\omega t - \frac{\pi}{2}\right) = -\cos(\omega t)$$

1.6 Double-Sideband Suppressed-Carrier (DSBSC) Modulation

In the standard form of amplitude modulation, the carrier wave $c(t)$ is completely independent of message signal $m(t)$, which means that the transmission of carrier wave represents a waste of power. This points to a shortcoming of amplitude modulation: namely that only a fraction of the total transmitted power is affected by $m(t)$. to overcome this shortcoming, we may suppress the carrier component from the modulated wave, resulting in double- sideband suppressed carrier modulation.

1.6.1 Time-Domain Description

$$s(t)_{DSB-FC} = A_C[\,1 + k_a m(t)]\,cos2\pi f_c t \tag{1.9}$$

$$= A_C\,cos2\pi f_c t + k_a m(t) A_C\,cos2\pi f_c t$$

$$s(t) = c(t).m(t)$$

$$= A_c cos2\pi f_c t.\,m(t)$$

When $k_a = 1$

$$s(t)_{DSB} = A_C m(t) cos2\pi f_c t \tag{1.10}$$

1.6.2 Frequency-Domain-Description

The suppression of the carrier from the modulated wave is well-appreciated by examining its spectrum. Specifically, by taking the Fourier transform

$$S(f) = \frac{1}{2}A_c[M(f-f_c) + M(f+f_c)] \tag{1.11}$$

where, as before, $S(f)$ is the Fourier transform of the modulated wave $s(t)$ and $M(f)$ is the Fourier transform of the message signal $m(t)$. When the message signal $m(t)$ is limited to the interval $-W \le f \le W$, except for a change in scale factor, the modulation process simply translates the spectrum of the baseband signal by $\pm f_{c'}$, of course, the transmission bandwidth required by DSBSC modulation is the same.

1.6.3 Generation-of-DSB-SC-Waves

A double-sideband suppressed-carrier modulated wave consists simply of the product of the message signal and the carrier wave. A device for achieving this requirement is called a product modulator. In this section, we describe two forms of a product modulator - the balanced modulator and the ring modulator.

Balanced Modulator

Two regular amplitude modulators are combined to create a balanced modulator, which suppresses the carrier wave. We consider the two modulators to be identical, with the exception of the sign change the modulating wave applied to one of them. Thus, the outputs of the two modulators may be as follows-

$$s(t)_1 = A_d[\,1 + k_a m(t)]\,cos2\pi f_c t$$

and

$$s(t)_2 = A_c[\,1 + k_a m(t)]\,cos2\pi f_c t$$

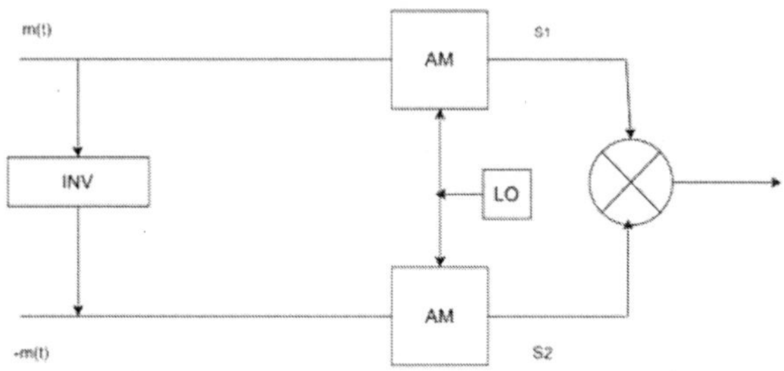

Fig. 1.13 Balanced Modulator

$$s(t) = s_1(t) - s_2(t)$$

$$s(t) \;=\; 2k_a A_c cos2\pi f_c t.m(t) \qquad (1.12)$$

1.6.4 Ring Modulator

One of the most useful product modulators that is well suited for generating a DSBSC modulated wave is the ring modulator, it is also known as a lattice or double-balanced modulator. The four diodes in Fig. 1.14 shown below forms ring in which they all point the same way. The diodes are controlled by a square-wave carrier $c(t)$ of frequency $f_{c'}$.Which is applied by means of two center-tapped transformers. We assume that the diodes are ideal, and the transformers are perfectly balanced. When the carrier supply is positive, the outer diodes (D_1, D_2) are switched on, presenting zero impedance, whereas the inner diodes (D_3, D_4) are switched off, presenting infinite impedance.

We can see that the modulator does not produce any output at the carrier frequency; instead, it only produces modulation products.

Operation of the circuit

- The operation is explained assuming that the diodes act as perfect switches and are turned on and off by an RF carrier signal. This is because the carrier's amplitude and frequency are greater than those of the modulating signal.
- For positive half cycle of $c(t)$ Diode D_1 and D_2 is on.
- For positive half cycle of $c(t)$ Diode D_3 and D_4 is on.

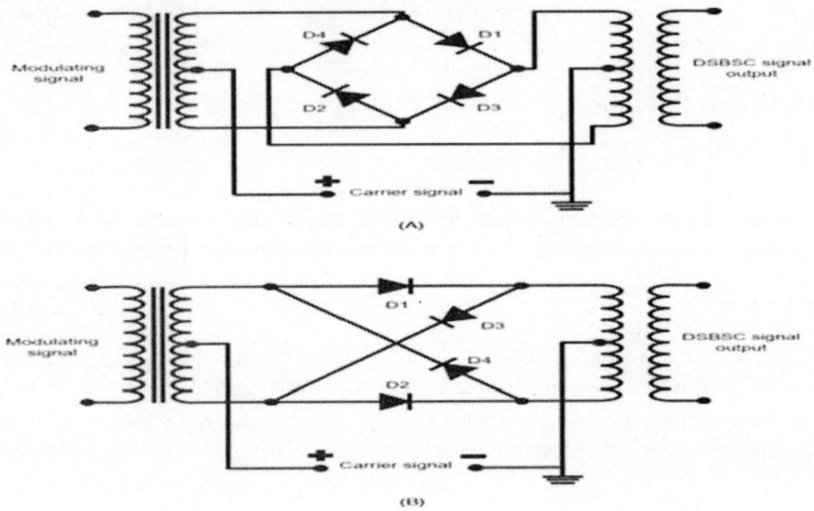

Fig. 1.14 Ring Modulator

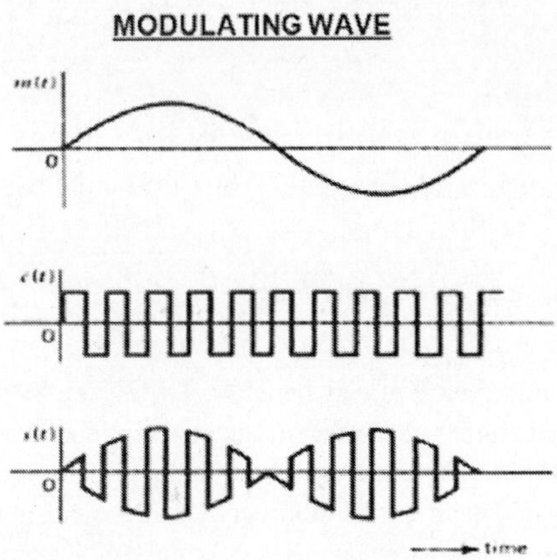

Fig. 1.15 Waveform for DSB

The square wave carrier $c(t)$ can be represented by a Fourier series-

$$c(t)\frac{4}{\pi}=\frac{(-1)^{n-1}}{2n-1}\cos\sum[2\pi f_c t(2n-1)] \tag{1.13}$$

Therefore, the ring Modulator output is

$$=\frac{4}{\pi}\frac{(-1)^{n-1}}{2n-1}\cos\sum[2\pi f_c t(2n-1)]\,m(t) \tag{1.14}$$

1.6.5 Coherent Detection of DSB-SC Modulated Wave

The message signal $m(t)$ is recovered from a DSBSC wave $s(t)$ by first multiplying $s(t)$ with a locally generated sinusoidal wave and then low-pass filtering the product. It is assumed that the local oscillator output is exactly coherent or synchronized, in both frequency and phase; with the carrier wave $c(t)$ used in the product modulator to generate $s(t)$. This method of demodulation is known as coherent detection or synchronous detection.

It is instructive to derive coherent detection as a special case of the more general demodulation process using a local oscillator signal of the same frequency but arbitrary phase difference ϕ, measured with respect to the carrier wave $c(t)$. Thus, denoting the local oscillator signal by $\cos(2\pi f_c t+\phi)$, assumed to be of unit amplitude for convenience, and for the DSBSC modulated wave $s(t)$, we find that the product modulator output.

$$\begin{aligned} v(t) &= \cos(2\pi f_c t+\phi)s(t) \\ &= A_c\cos(2\pi f_c t)\cos(2\pi f_c t+\phi)m(t) \\ &= \frac{1}{2}A_c\cos\phi m(t)+\frac{1}{2}A_c\cos(4\pi f_c t+\phi)m(t) \end{aligned} \tag{1.15}$$

$$\begin{aligned} v(t) &= \cos(2\pi f_c t+\phi)s(t) \\ &= A_c\cos(2\pi f_c t)\cos(2\pi f_c t+\phi)m(t) \end{aligned}$$

$$=\frac{1}{2}A_c\cos\phi m(t)+\frac{1}{2}A_c\cos(4\pi f_c t+\phi)m(t) \tag{1.16}$$

Now, $v(t)$ is passed through a low-pass filter. Thus, the output become

$$y(t)=\frac{1}{2}A_c\cos\phi m(t) \tag{1.17}$$

Maximum when $\phi=0$, and is minimum (zero) when $\phi=\pm\pi/2$. The zero demodulated signal, which occurs for $\phi=\pm\pi/2$, represents the quadrature null effect of the coherent detector. Thus, the phase error ϕ in the loca oscillator causes the detector output to be attenuated by a factor equal to $\cos\phi$. As long as the phase error ϕ is constant, the detector output provides an undistorted version of the original message signal $m(t)$. In practice, however, we usually find that the phase error ϕ varies randomly with time, owing to random variations in the communication channel. The result is that at the detector output, the multiplying factor cosh also varies randomly with time, which is obviously undesirable. Therefore, circuitry must be provided in the receiver to maintain the local oscillator in perfect synchronism, in both frequency and phase, with the carrier wave used to generate the

DSBSC modulated wave in the transmitter. The resulting increase in receiver complexity is the price that must be paid for suppressing the carrier wave to save transmitter power.

1.6.6 Costas Loop

One method of obtaining a practical synchronous receiving system, suitable for use with DSBSC modulated waves, is to use the Costas Loop. It consists of two coherent detectors supplied with the same input signal namely, the incoming DSBSC modulated wave $A_c\cos(2\pi f_c t)$, but with individual local oscillator signals that are phase quadrature to each other. The frequency of the local oscillator is adjusted to be the same as the carrier frequencyf_c, which is assumed known a priori. The detector in the upper path is referred to as the in phase coherent detector or l-channel, and that in the lower path is referred to as the quadrature-phase coherent detect or Q-channel. These two detectors are coupled to form a negative feedback system designed in such a wayb to maintain the local oscillator synchronous with the carrier wave.

1.6.7 Coherent (Synchronous) Detection of DSB-SC Waves

- The coherent detector for the DSB-SC signal is shown in Fig. 1.16.
- The DSB-SC Wave $s(t)$ is applied to a product modulator in which it is multiplied with the locally generated carrier $\cos(2\pi f_c t)$

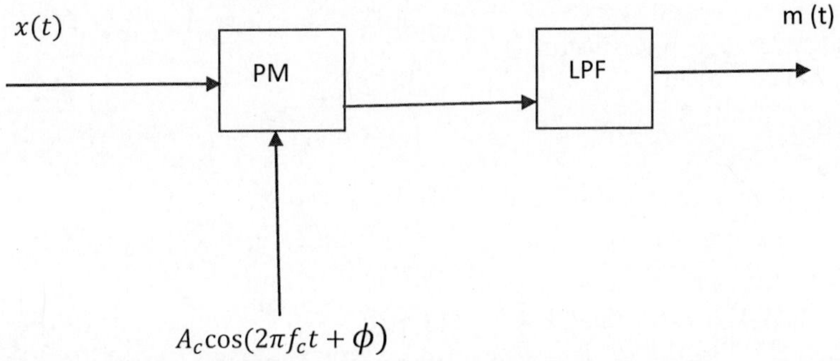

Fig.1.16 Product Modulator

Analysis of coherent Detection:

- Let $x(t)$ be the DSB-SC signal at the input of the product modulator and the local oscillator having frequency$A_c\cos(2\pi f_c t + \phi)$. The signal $x(t)$ can be represented as-

$$y(t) = \frac{1}{2}A_c \cos\phi m(t)\, x(t) = m(t) \times A_c \cos(2\pi f_c t) \quad (1.18)$$

Hence output of the product modulator is given by

$$x'(t) = m(t) \cdot A_c\cos(2\pi f_c t)\cos(2\pi f_c t + \phi)$$
$$x'(t) = m(t) \cdot A_c\cos(2\pi f_c t + \phi)\cos(2\pi f_c t)$$

$$\cos A \cos B = \frac{1}{2}[\cos(A+B) + \cos(A-B)]$$
$$x'(t) = \frac{1}{2}m(t)A_c[\cos(4\pi f_c t + \phi) + \cos\phi]$$

$$x'(t) = \frac{1}{2}A_c \cos\phi m(t) + \frac{1}{2}m(t)A_c \cos(4\pi f_c t + \phi) \quad (1.19)$$

Signal $x'(t)$ is then passed through a low pass filter. Which allows only the first term to pass through and will reject the second term. Hence the filter output is given by,

$$m'(t) = \frac{1}{2}A_c \cos\phi m(t) \quad (1.20)$$

Advantages

- Power is not wasted in the form of carrier
- SB power increases.
- Efficiency is 100 %.
- SNR is improved and compared to AM.

Dis- Advantages:

- Complexity of receiver is very high
- Transmission Bandwidth is $2f_m$.

Application

- Use in point-to-point communication in short distances
- It is used for carrying audio.

1.7 SINGLE SIDEBAND MODULATION (SSB)

The number of channels that can be accommodated in a given frequency space increases when the modulated signal's bandwidth is reduced because the message signal appears twice in a DSB-SC signal, making it unnecessary to increase the bandwidth. In order to accommodate twice as many channels in a given frequency space, using single side one in place of both sidebands is known as single sideband suppressed carrier modulation of this type, which provides a single sideband with suppressed carrier and contains all of the information content in the message signal once and simultaneously reduces the bandwidth by half (SSB-SC). The expression of AM modulated signal can be represented as,

$$s(t) = A_c(1 + k_a m(t)]cos\omega_c t \quad (1.21)$$

where $m(t)$ is message signal $A_c cos\omega_c t$ is carrier signal and k_a is modulation coefficient of modulator.

In Fourier transform domain modulated signal can be represented as,

$$S(\omega) = A_c\pi\delta(\omega_c + \omega) + A_c\pi\delta(\omega - \omega_c) + \frac{A_c}{2} k_a M(\omega_c + \omega) + \frac{A_c}{2} k_a M(\omega - \omega_c). \quad (1.22)$$

Because they are complex conjugates of one another, the spectral components of the AM signal at equal intervals above and below the carrier frequency carry the same information.

$$a(t) = A_c m(t) cos\omega_c t \tag{1.23}$$

$$A(\omega) = \frac{A_c}{2} M(\omega_c + \omega) + \frac{A_c}{2} M(\omega - \omega_c) \tag{1.24}$$

and is centred at the carrier frequency ω_c. Then $H(\omega)$ selects the desired sideband. Upper sideband modulation uses the high pass filter

$$H_u(\omega) = \begin{cases} 1 \; for \; |\omega| > \omega_c \\ 0 \quad elsewhere \end{cases} \tag{1.25}$$

and the lower sideband SSB modulation uses the lowpass filter

$$H_l(\omega) = \begin{cases} 1 \; for \; |\omega| < \omega_c \\ 0 \quad elsewhere \end{cases} \tag{1.26}$$

let the baseband message be $m(t)$ and its Hilbert transform $\widehat{m}(t)$the pre-envelope of SSB signal has the transform

$$\begin{aligned} S_+(\omega) &= 2S(\omega)u(\omega) \\ &= 2A(\omega)H(\omega)u(\omega) \\ &= A_c M(\omega - \omega_c)H(\omega)u(\omega) \end{aligned} \tag{1.27}$$

And the transform of its complex envelope is

$$\tilde{S}(\omega) = S_+(\omega_c + \omega) = A_c M(\omega)H(\omega_c + \omega)u(\omega_c + \omega) \tag{1.28}$$

Upper Sideband case

Substituting $H_u(\omega)$ for $H(\omega)$ gives

$$\tilde{S}(\omega) = A_c M(\omega)u(\omega) = \frac{A_c}{2} M(\omega)(1 + sign\omega) \tag{1.29}$$

The complex envelope is

$$\frac{A_c}{2}[m(t) + j\widehat{m}(t)] \tag{1.30}$$

So, SSB signal can be represented as

$$s(t) = Re\{\tilde{s}(t)e^{j\omega t}\} \tag{1.31}$$

$$= \frac{A_c}{2} Re\{[m(t) + j\hat{m}(t)]e^{j\omega t}$$
$$= \frac{A_c}{2} m(t)cos\omega_c t - \frac{A_c}{2}\hat{m}(t)sin\omega_c t$$

(1.32)

For **lower sideband case** the corresponding SSB signal can be given as

$$s(t) = \frac{A_c}{2} m(t)cos\omega_c t + \frac{A_c}{2}\hat{m}(t)sin\omega_c t$$

where + and – signs correspond to the lower sideband and upper sidebands respectively.

1.7.1 Generation of SSB-SC Signals

SSB-SC modulated signal can be generated by two methods

i) Filter method or frequency discrimination method
ii) Phase discrimination method

Filter Method

In this method first DSB-SC signal $s_{DSB-SC}(t)$ is generated by using a simple product modulator or a balanced modulator that gives the product of message signal $m(t)$ and carrier signal $A_C cos\omega_C t$, and then one of the sidebands is filtered out by appropriate bandpass filter $H(\omega)$. The diagram is shown in Fig. 1.17, the design of bandpass filter is very critical and put some limitation on the modulating and carrier frequencies.

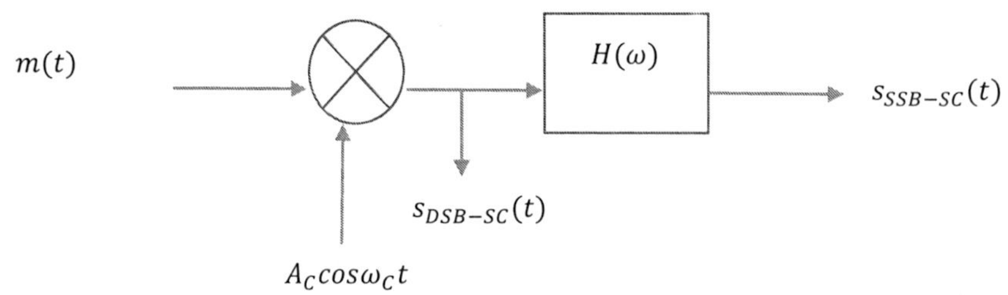

Fig.1.17 Filter method for generating SSB-SC

Phase discrimination method

The block diagram of phasing method is given in Fig. 1.18, the product term $\frac{A_c}{2} m(t)cos\omega_c t$ is an AM-SC signal which can be generated by simple product modulator or balanced modulator. The product term $\frac{A_c}{2}\hat{m}(t)sin\omega_c t$ is generated by phasing $\hat{m}(t)$ and $sin\omega_c t$ through another product or balance modulator. The function $\hat{m}(t)$ generated by passing $m(t)$ through a wideband $(-\frac{\pi}{2})$ phase-shifter. A wide band-phase shifter is needed so that all the frequency component present in baseband signal is shifted by (90°) keeping their amplitude unchanged. Signal $sin\omega_c t$ obtained by passing $cos\omega_c t$ through a simple (-90°) phase shifter. The product terms $\frac{A_c}{2}m(t)cos\omega_c t$ and $\frac{A_c}{2}\hat{m}(t)sin\omega_c t$ are added by adder to generate SSB-SC signal

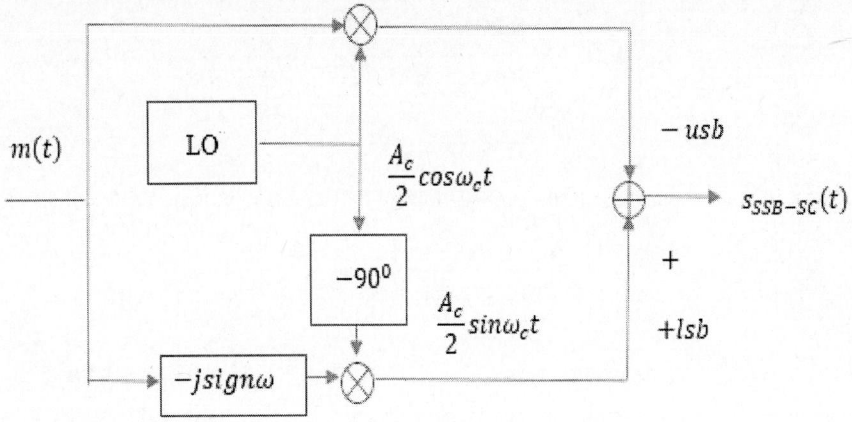

Fig. 1.18 Phase shift method for SSB-SC generation

Signal Fourier Transforms in Steps for Generating an Upper-Sideband SSB Signal

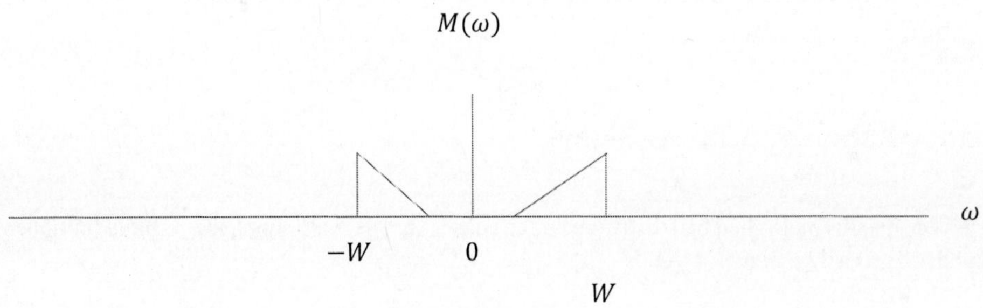

Fig. 1.19 Fourier transform of the baseband message

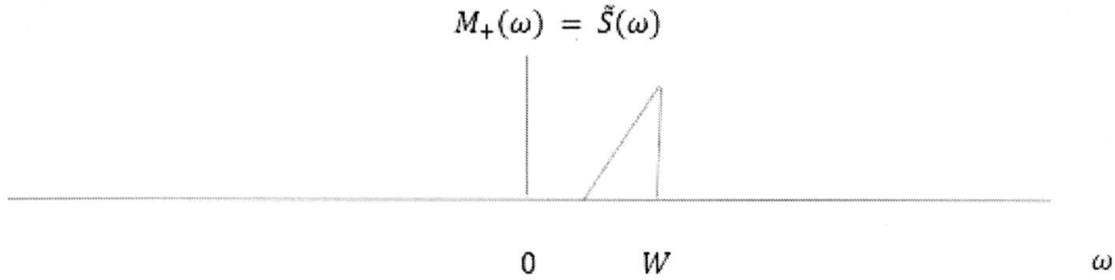

Fig. 1.20 Fourier transform of the pre-envelope of the baseband message

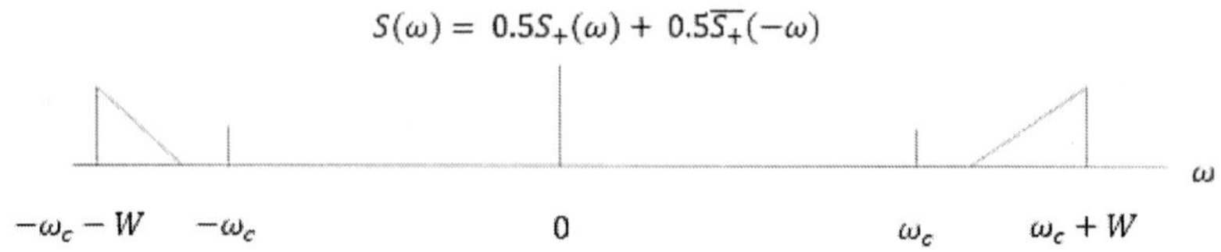

Fig. 1.21 Fourier transform of the upper-sideband transmitted signal

1.7.2 Demodulation of SSB-SC Signal

First the received signal is multiplied by locally generated carrier signals multiplying the formula for upper and lower sideband SSB signal by $2cos\omega_c t$ yields

$$b(t) = A_c m(t)cos^2\omega_c t \mp A_c \hat{m}(t)sin\omega_c t cos\omega_c t \tag{1.33}$$
$$= \frac{A_c}{2} m(t) + \frac{A_c}{2} m(t)cos2\omega_c t \mp \frac{A_c}{2} \hat{m}(t)sin2\omega_c t$$

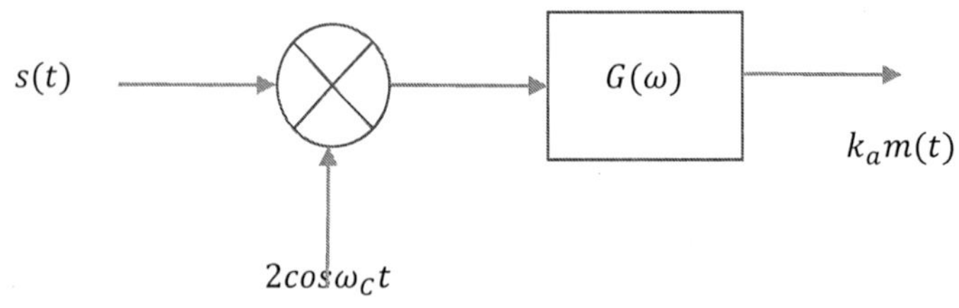

Fig. 1.22 SSB-SC demodulator

Here, $\frac{A_c}{2}\, m(t)$ is the desired component, and $\frac{A_c}{2}\, m(t)cos2\omega_c t$ and $\frac{A_c}{2}\, \hat{m}(t)sin2\omega_c t$ have spectra centered about $2\omega_c$, that will be removed buy low pass filter $G(\omega)$ with cut off frequency W.

Example 1.6: - which of the following is a single tone SSB modulation

a) $cos\omega_C t$
b) $cos\omega_C t + cos\omega_m t$
c) $cos(\omega_C + \omega_m)t$
d) None of the above

1.8 VSB MODULATION

As discussed earlier, it is rather difficult to generate exact SSB signals. They generally require that message signal $m(t)$ have a null around dc. A phase shifter is required in the phase shift method is unrealizable or approximately realizable. The generation of DSB signals is much simpler but requires twice the signal bandwidth. Vestigial Side Band (VSB) modulation is called the asymmetric sideband system, is a compromise between DSB and SSB. It inherits the advantageous of DSB and SSB but avoids their disadvantageous at the small cost. VSB signals are relatively easy to generate and at the same time their bandwidth is only a little (typically 25%) greater than that of SSB signals.

1.8.1 VSB Modulator

A VSB-SC signal can be generated by passing DSB-SC signal through an appropriate filter. The block diagram of VSB modulator is shown in Fig. 1.23. The modulating signal $m(t)$ and the carrier $A_c cos\omega_c t$ is applied to the product modulator. The output of the product modulator in time domain is given

$$S_{DSB-SC}(t) = A_c m(t) cos\omega_c t \tag{1.34}$$

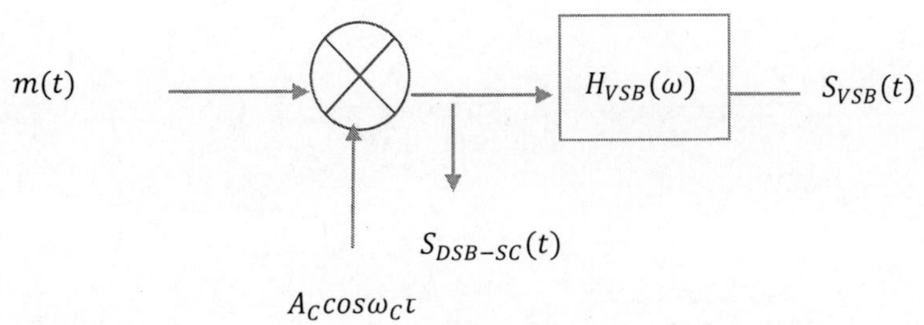

Fig. 1.23 VSB Modulator

This represents a DSB-SC modulated wave. This DSB-SC signal is then applied to a sideband shaping filter $H_{VSB}(\omega)$. The design of this shaping filter depends on the desired spectrum of the VSB modulated signal. This filter will pass the required sideband (usually USB) and the vestige of the other (LSB) sideband.

1.8.2 Spectrum of a VSB signal

Suppose a message signal $m(t)$, and in transform domain it is represented by $M(\omega)$, has maximum frequency W, carrier frequency ω_C , and small vestige x then the spectra of message signal, DSB-SC signal and VSB-SC signal can be represented as follows

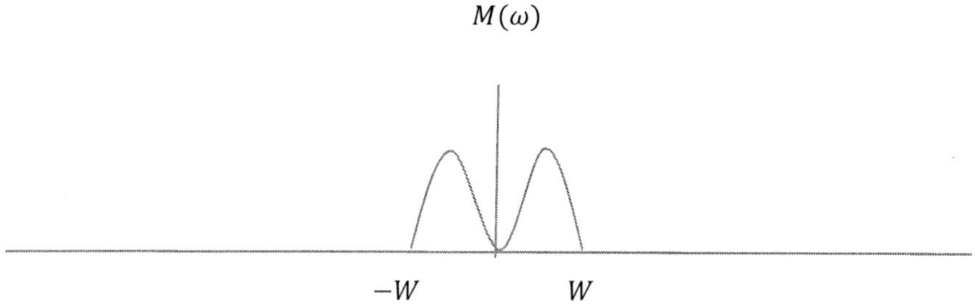

Fig. 1.24 Spectra of modulating signal

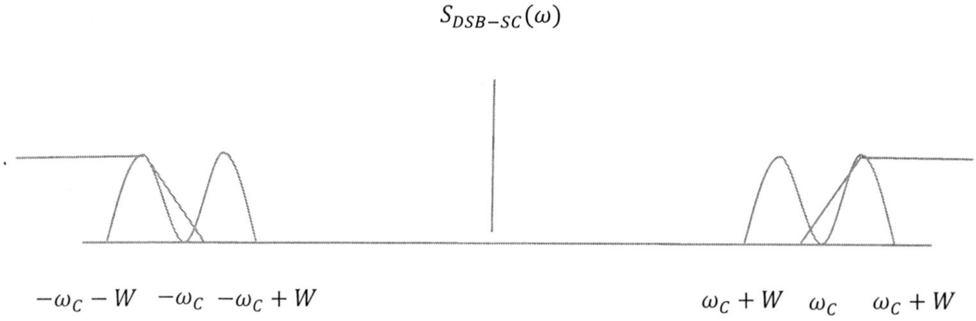

Fig. 1.25 Spectra of double side band signal

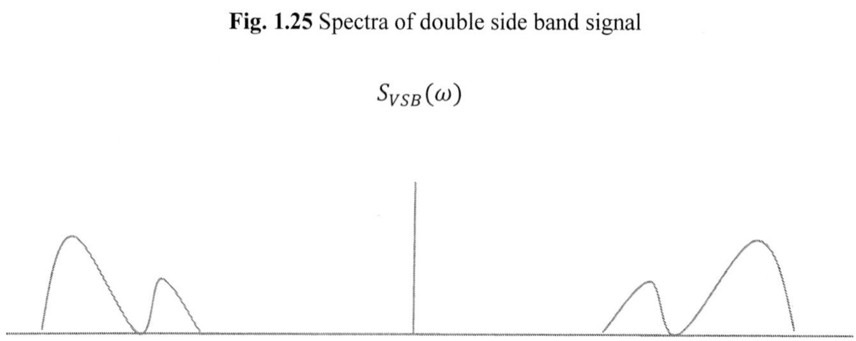

$\omega_C - x$ ω_C $\omega_C + W$

$-\omega_C - W$ $-\omega_C$ $-\omega_C + x$

Fig. 1.26 Spectra of VSB signal

In the above spectrum, we consider a system in which the lower sideband contributes the vestigial sideband portion while the upper sideband remains the major band. However, either sideband may be used to create the main band.

1.8.3 VSB Filter Characteristics

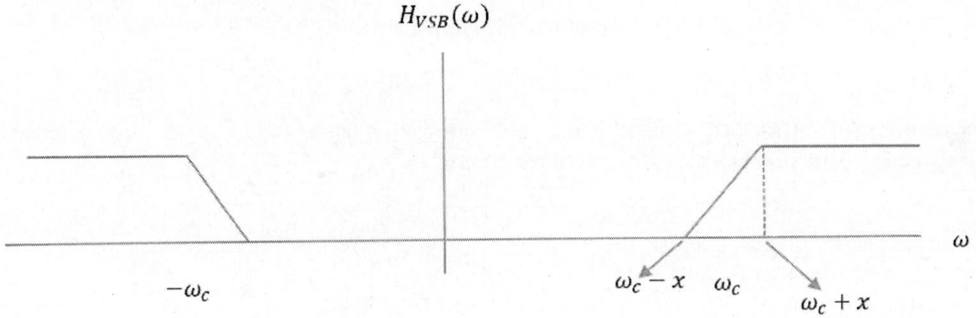

Fig. 1.27 VSB filter characteristics

In the Fig. 1.27 given, the filter has a gain of K for all frequencies $|\omega| > \omega_c + x$ (passband). For the frequencies $\omega_c < |\omega| < \omega_c + x$, which reside in the principal band, the gain is less than K, so some loss compared to the passband occurs. For the frequencies $\omega_c - x < |\omega| < \omega_c$, the filter's gain is not 0, therefore some of the other sideband's frequency components are passed by the filter and $S_{VSB}(t)$ has a vestigial sideband. The bandwidth of $S_{VSB}(t) = W + x$. Generally, since we are trying to reduce the bandwidth compared to DSB-SC, x is smaller than W.

1.8.4 Filter characteristics For VSB modulation

Characteristics of filter in transform domain can be given as

$$S_{VSB}(\omega) = S_{DSB-SC}(\omega)\, H_{VSB}(\omega) \quad (1.35)$$

$$= \frac{A_c}{2} M(\omega - \omega_c) H_{VSB}(\omega) + \frac{A_c}{2} M(\omega + \omega_c) H_{VSB}(\omega)$$

$$= \frac{A_c}{4} M_+(\omega - \omega_c) H^+_{VSB}(\omega) + \frac{A_c}{4} M_-(\omega - \omega_c) H^+_{VSB}(\omega)$$

$$+\frac{A_c}{4}M_+(\omega+\omega_c)H^-_{VSB}(\omega)+\frac{A_c}{4}M_-(\omega+\omega_c)H^-_{VSB}(\omega)$$

where

$$H^+_{VSB}(\omega)=\begin{cases}H_{VSB}(\omega) & \omega>0\\ 0 & \omega<0\end{cases} \quad \text{and} \quad H^-_{VSB}(\omega)=\begin{cases}H_{VSB}(\omega) & \omega<0\\ 0 & \omega>0\end{cases}$$

We note that $H^-_{VSB}(\omega)=H^{+*}_{VSB}(\omega)$ due to Hermitian symmetry in the frequency response of real systems.

1.8.5 Demodulation of VSB

The block diagram of synchronous detector of VSB modulated wave is show in Fig. 1.28. The VSB modulated wave is passed through a product modulator, where it is multiplied with the locally generated carrier. The output of product modulator can be given as,

$$x(t) = A_r S_{VSB}(t) cos\omega_c t$$

If we want $z(t)=Gm(t)$, then we need to impose a constraint on the frequency response of the modulators filter $H_{VSB}(\omega)$. In transform domain $x(t)$ can be represented as

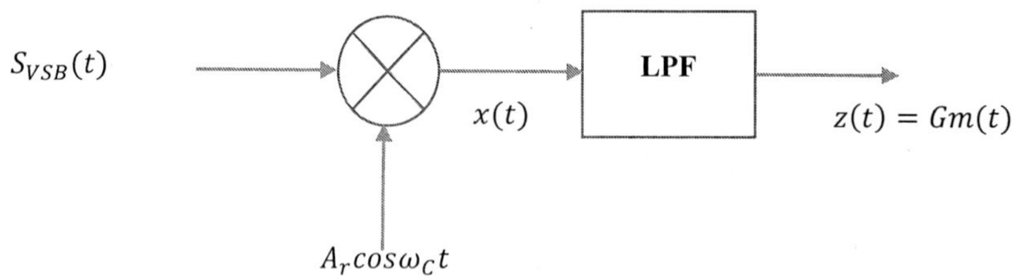

Fig. 1.28 VSB demodulator

$$X(\omega)=\frac{A_r}{2}S_{VSB}(\omega-\omega_c)+\frac{A_r}{2}S_{VSB}(\omega+\omega_c) \tag{1.36}$$

$$=\frac{A_cA_r}{8}M_+(\omega-2\omega_c)H^+_{VSB}(\omega-\omega_c)+\frac{A_cA_r}{8}M_-(\omega-2\omega_c)H^+_{VSB}(\omega-\omega_c)$$

$$+\frac{A_cA_r}{8}M_+(\omega)H^-_{VSB}(\omega-\omega_c)+\frac{A_cA_r}{8}M_-(\omega)H^-_{VSB}(\omega-\omega_c)$$

$$+\frac{A_cA_r}{8}M_+(\omega)H^+_{VSB}(\omega+\omega_c)+\frac{A_cA_r}{8}M_-(\omega)H^+_{VSB}(\omega+\omega_c)$$

$$+\frac{A_cA_r}{8}M_+(\omega+2\omega_c)H^-_{VSB}(\omega+\omega_c)+\frac{A_cA_r}{8}M_-(\omega+2\omega_c)H^-_{VSB}(\omega+\omega_c)$$

$$Z(\omega)=\frac{A_cA_r}{8}M_+(\omega)H^-_{VSB}(\omega-\omega_c)+\frac{A_cA_r}{8}M_-(\omega)H^-_{VSB}(\omega-\omega_c) \quad (1.37)$$

$$+\frac{A_cA_r}{8}M_+(\omega)H^+_{VSB}(\omega+\omega_c)+\frac{A_cA_r}{8}M_-(\omega)H^+_{VSB}(\omega+\omega_c)$$

$$Z(\omega)=\frac{A_cA_r}{8}M_+(\omega)\{H^-_{VSB}(\omega-\omega_c)+H^+_{VSB}(\omega+\omega_c)\}$$

$$+\frac{A_cA_r}{8}M_-(\omega)\{H^-_{VSB}(\omega-\omega_c)+H^+_{VSB}(\omega+\omega_c)\}$$

We want $Z(\omega)=GM(\omega)$, where G is a constant. If we ensure that

$$H^-_{VSB}(\omega-\omega_c)+H^+_{VSB}(\omega+\omega_c)=K$$

$$Z(\omega)=\frac{A_cA_rK}{8}M_+(\omega)+\frac{A_cA_rK}{8}M_-(\omega)=\frac{A_cA_rK}{4}M(\omega)$$

Therefore $z(t)=\frac{A_cA_rK}{4}m(t)$. If we replace $H^-_{VSB}(\omega)=H^{+^*}_{VSB}(\omega)$ and ω by $\Delta\omega$ then we get

$$H^{+^*}_{VSB}(\omega_c-\Delta\omega)+H^+_{VSB}(\omega_c+\Delta\omega)=K \quad (1.38)$$

These criteria need to be true over the frequency range of the VSB signal.

Possible VSB filters are filter with linear transition bands and raised cosine filters.

Question: The signal $m(t) = 2cos(2\pi10t)+3cos(2\pi30t)$. We wish to transmit this signal using VSB with carrier $c(t) = 5cos\,(2\pi500t)$. The VSB filter's response is shown below. Find $S_{VSB}(t)$ as well as its bandwidth.

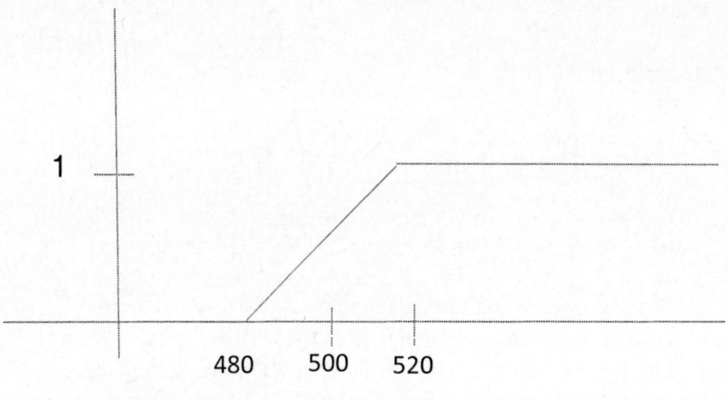

1.9 ANGLE MODULATION

Angle modulation is another significant modulation technique used for message transmission, where the carrier wave phase angle is varied in accordance with the modulating signal. An important advantage of angle modulation is noise immunity over amplitude modulation. However, this advantage against noise is at the cost of increased bandwidth.

1.9.1 Basic Definition

In angle modulation, the phase angle of a carrier is varying accordance with the instantaneous value of the modulating signal. Let us consider the angle modulated wave as

$$x_c(t) = A\ \cos[\omega_c t + \phi(t)] \tag{1.39}$$

where, A is the amplitude of carrier, ω_c is the carrier frequency and $\phi(t)$ is some phase angle.
Let ψ denotes the total phase angle of the carrier wave, assumed to be the function of message signal is given as

$$\psi = \omega_c t + \phi(t) \tag{1.40}$$

Eq. (1.37) can be considered as the real part of a rotating phaser $Ae^{j\psi}$. Let it be denoted by $\hat{\psi}$,

$$\hat{\psi} = Ae^{j\psi} \tag{1.41}$$

$$= Re[Ae^{j\psi}]$$
$$= A\cos\psi$$

From Eq. (1.38), the constant angular velocity ω_c of the phaser $\hat{\psi}$ is related to the total phase angle $\psi = \omega_c t + \phi(t)$
Differentiating the above equation both side with respect to *t*, we have

$$\omega_i = \frac{d\psi}{dt} \tag{1.42}$$
$$= \omega_c + \frac{d\phi(t)}{dt}$$

when $\phi(t)$ is constant, then

$$\frac{d\psi}{dt} = \omega_c \tag{1.43}$$

the derivative in general vary with time. This time dependent angular frequency is called instantaneous angular frequency which provides a time varying instantaneous frequency f_i of the carrier.
If the angle ψ is made to vary in accordance with the instantaneous value of the modulating signal, the carrier is said to be angle modulation. Based on phase angle variation, the two basic type of angle modulation are phase modulation (PM) and frequency modulation (FM)

1.9.2 Representation of PM and FM Signals

In PM, the instantaneous phase deviation of the carrier is proportional to the message signal ($m(t)$), is given as

$$\phi(t) = k_p m(t) \tag{1.44}$$

Where k_p is the proportionality constant known as phase sensitivity of the modulation, expressed in $\frac{radian}{volts}$.

The instantaneous value of phase angle is given

$$\psi_i = \omega_c t + k_p m(t) \tag{1.45}$$

Thus the expression of phase modulated wave is given as

$$\begin{aligned} x_p(t) &= A \cos \psi_i \\ &= A \cos[\omega_c t + k_p m(t)] \end{aligned} \tag{1.46}$$

Eq. (1.46) is the required mathematical expression for a phase modulated wave.

In FM, the instantaneous frequency is varied with the message signal. The instantaneous value of angular frequency ω_i for frequency modulation is given as

$$\omega_i = \omega_c + k_f m(t) \tag{1.47}$$

where, k_f is the proportionality constant known as frequency sensitivity of the modulated wave. k_f is expressed in $\frac{Hz}{volt}$.

From Eq. (1.42), we know that

$$\omega_i = \frac{d\psi_i}{dt}$$

the instantaneous phase angle is given as

$$\psi_i = \int \omega_i \, dt \tag{1.48}$$

Sunbsituting the value of $\omega_i$in Eq. (1.46) from Eq. (1.45), we get

$$\begin{aligned} \psi_i &= \int [\omega_c + k_f m(t)] dt \\ &= \omega_c t + k_f \int m(t) dt \end{aligned} \tag{1.49}$$

Thus, we can express the frequency modulated signal as

$$x_f(t) = A \cos \psi_i \tag{1.50}$$

$$= A\cos[\omega_c t + k_f \int_0^t m(t)\,dt]$$

We know that instantaneous frequency of FM is given as

$$\omega_i = \omega_c + k_f m(t) \tag{1.51}$$

The instantaneous frequency varies around the carrier frequency ω_c, The maximum change in the instantaneous frequency from the ω_c is called frequency deviation. The $\Delta\omega$ is defined as

$$\Delta\omega = |\omega_i - \omega_c|_{max} \tag{1.52}$$

is called the maximum radian frequency deviation of angel modulated signal.

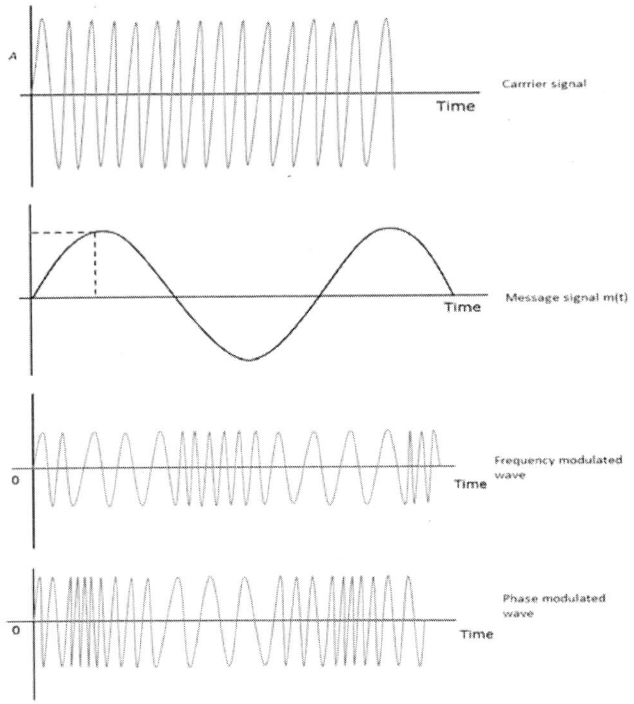

Fig. 1.29 FM and PM waveform

Example 1.7: A FM signal is represented by the following equation

$$x(t) = 15\cos(2\times 10^4 t + 5\sin 1050t)$$

Determine the following

1. **Carrier frequency**
2. **Modulating frequency**
3. **Maximum deviation**

Solution:

The standard expression for a FM wave is given as

$$x(t) = A\cos(w_c t + k_f \int x(t)dt)$$

The given equation is

$$x(t) = 15\cos(2 \times 10^4 t + 5\sin 1050t)$$

Comparing both the equation, we get

1. Carrier frequency

$$\omega_c = 2 \times 10^4 rad/sec.$$

$$f_c = \frac{2 \times 10^4}{2\pi} = 3.18MHz$$

2. Modulation frequency

$$\omega_m = 1050\, rad/sec$$

$$f_m = \frac{1050}{2\,\pi} = 167.11Hz$$

3. Maximum frequency deviation

$$\Delta\omega = 5 \times f_m = 5 \times 167.11$$
$$= 825.5\ Hz$$

Example 1.8: - Determine the frequency deviation for a frequency modulated signal which has resting frequency of 100 MHz and upper frequency is 100.01MHz when modulated. Find the lowest frequency reached by the FM wave.

Solution:

Given that

Carrier frequency $f_c = 100MHz$

$f_h = 100.01MHz$

We know that frequency deviation $\Delta_f = f_h - f_c$

$$= 100 - 100.01$$
$$= 0.01MHz$$

Lowest frequency reached by the modulated wave is given as

$$f_l = f_c - \Delta f$$
$$= 100 - 0.01$$
$$= 99.99\ MHz$$

1.10 Types of FM

The FM signal depends on the frequency deviation $\Delta\omega = k_f m(t)$. Thus, based on frequency deviation the bandwidth of the FM signal can be large or small. Since $\Delta\omega = k_f m(t)$, hence, when k_f is small then the bandwidth will be narrow and when k_f is large the bandwidth is wide.

i). **Narrowband FM**: When k_f is small, the bandwidth of FM is narrow.

ii). **Wideband FM:** When k_f has an appreciable value, then FM signal has wide bandwidth.

1.10.1 Spectral Characteristics of Angle Modulated Signals

As is mentioned in previous section 1.7, the modulated carrier for angle modulation is represented by

$$x_c(t) = A\ \cos[\omega_c t + \phi(t)]$$

where A and ω_c are constants and the phase angle $\phi(t)$ is a function of the message signal $m(t)$. The angle modulated carrier can be represented in the exponential form as

$$x_c(t) = \mathrm{Re}\left(A\ \mathrm{e}^{\mathrm{j}(\omega_c t + \phi(t))}\right) = \mathrm{Re}\left(A\ \mathrm{e}^{\mathrm{j}\omega_c t}\ \mathrm{e}^{\mathrm{j}\phi(t)}\right) \tag{1.53}$$

where Re(.) denotes the real part of (.).
Expanding $\mathrm{e}^{\mathrm{j}\phi(t)}$ in power series would give us

$$\begin{aligned} x_c(t) &= \mathrm{Re}\left\{Ae^{j\omega_c t}\left[1 + j\phi(t) - \frac{\phi^2(t)}{2!} - \cdots + j^n\frac{\phi^n(t)}{n!} + \cdots\right]\right\} \\ &= A\left[\cos\omega_c t - \phi(t)\sin\omega_c t - \frac{\phi^2(t)}{2!}\cos\omega_c t + \frac{\phi^3(t)}{3!}\sin\omega_c t + \cdots\right] \end{aligned} \tag{1.54}$$

From Eq. (1.54) we can observe that the angle modulated signal is comprised of unmodulated carrier and various amplitude modulated terms, which are $\phi(t)\sin\omega_c t$, $\phi^2(t)\cos\omega_c t$, $\phi^3(t)\sin\omega_c t$, ..., and so on. Or we can say that, the Fourier spectra of angle modulated signal consists of an unmodulated carrier and spectra of $\phi(t), \phi^2(t), \phi^3(t)$, ..., and so on, centred at ω_c.
It is evident that due to inherent nonlinearity of angle-modulation, its Fourier spectrum is difficult to characterise in a simple way, as was in the case of amplitude modulation. We will consider sinusoidal message signal $m(t)$ to observe the spectral properties of angle modulated signal

1.10.2 Angle Modulation by a Sinusoidal Signal

From the previous section (1.7), for both PM and FM

$$x_c(t) = A\ \cos[\omega_c t + \beta\sin\omega_m t] \tag{1.55}$$

Eq. (1.55) can be expressed as

$$x_c(t) = \mathrm{Re}\left(A\ \mathrm{e}^{\mathrm{j}\omega_c t}\ \mathrm{e}^{\mathrm{j}\beta\sin\omega_m t}\right) \tag{1.56}$$

As we know that the function $\mathrm{e}^{\mathrm{j}\beta\sin\omega_m t}$ is a periodic function with period $T_m = \frac{2\pi}{\omega_m}$. Thus, its Fourier series expansion can be represented as

$$\mathrm{e}^{\mathrm{j}\beta\sin\omega_m t} = \sum_{n=-\infty}^{\infty} c_n \mathrm{e}^{-\mathrm{j}n\omega_m t} \tag{1.57}$$

From the basics of Fourier series, the Fourier coefficients can be computed as

$$c_n = \frac{\omega_m}{2\pi}\int_{-\pi/\omega_m}^{-\pi/\omega_m} \mathrm{e}^{\mathrm{j}\beta\sin\omega_m t}\mathrm{e}^{-\mathrm{j}n\omega_m t}dt$$

Assume $\omega_m t = p$, therefore, $dt = \frac{dp}{\omega_m}$

$$c_n = \frac{1}{2\pi}\int_{-\pi/\omega_m}^{-\pi/\omega_m} \mathrm{e}^{\mathrm{j}(\beta\sin p - np)}dp = J_n(\beta)$$

Thus, we can write

$$e^{j\beta \sin \omega_m t} = \sum_{n=-\infty}^{\infty} J_n(\beta) e^{jn\omega_m t} \tag{1.58}$$

Substituting Eq. (1.58) in Eq. (1.56), we get

$$\begin{aligned} x_c(t) &= A \operatorname{Re}\left[e^{j\omega_c t} \sum_{n=-\infty}^{\infty} J_n(\beta) e^{jn\omega_m t} \right] \\ &= A \operatorname{Re}\left[\sum_{n=-\infty}^{\infty} J_n(\beta) e^{j(\omega_c + n\omega_m)t} \right] \end{aligned} \tag{1.59}$$

Extracting the real part, we finally get

$$x_c(t) = A \sum_{n=-\infty}^{\infty} J_n(\beta) \cos(\omega_c + n\omega_m)t \tag{1.60}$$

where $J_n(\beta)$ is the Bessel function of the first kind of order n and argument β. Using the values of $J_n(\beta)$ given in Appendix I, we can make following observations

1. The spectrum is comprised of carrier frequency component and other infinite number of sideband components at frequencies $\omega_c \pm n\omega_m$ as shown in Fig. 1.29.
2. The amplitude of the spectral lines depends on the value of $J_n(\beta)$.The value of $J_n(\beta)$ is very small for large value of n.
3. The number of significant spectral lines is a function of β. For $\beta \ll 1$, only J_0 and $J_1$are significant, therefore spectrum consist of carrier and two sidebands as shown in Fig. 1.29. However, if $\beta \gg 1$ there are many sidebands.

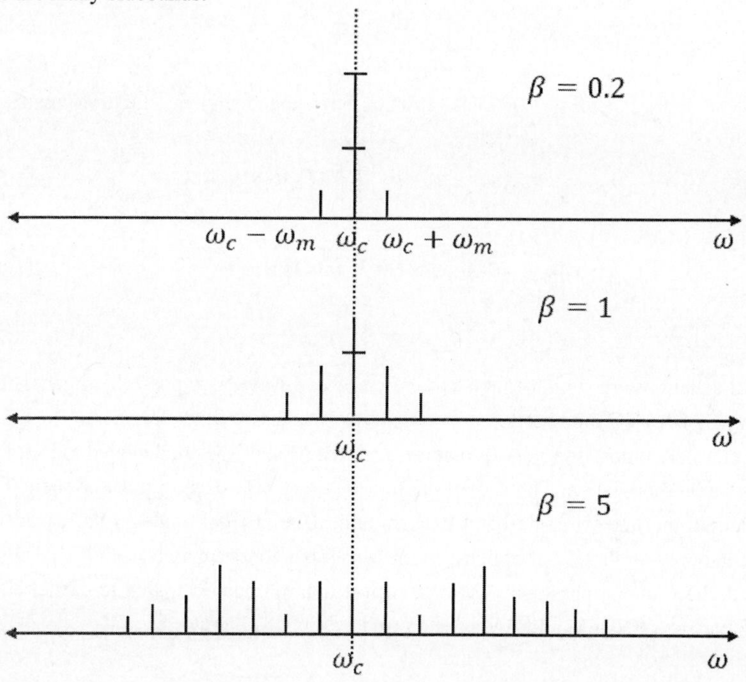

Fig. 1.29 Amplitude spectrum of sinusoidally modulated FM signals keeping ω_m fixed

1.11 Generation of Angle modulation

1.11.1 Narrowband angel modulated signal

The narrowband FM and PM are easily generated from the Eq. (1.61) and Eq. (1.62) Fig. (1.30) illustrate the generation of NBFM and NBPM

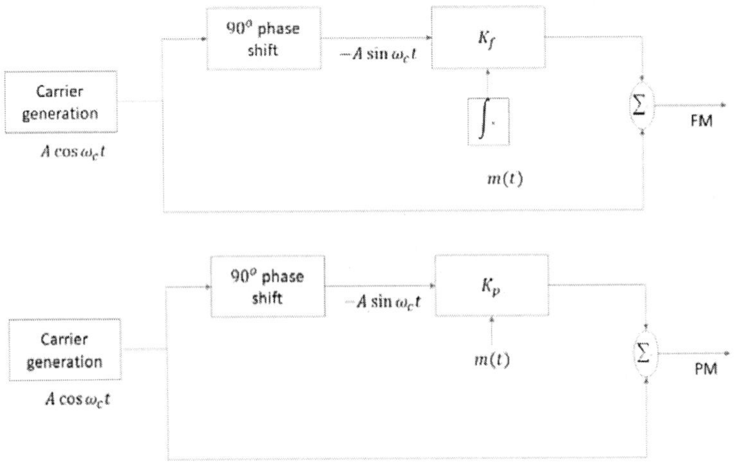

Fig. 1.30 Generation of narrowband angle modulation signals

$$x_{NBFM} \approx A[cos\omega_c t - k_f\left[\int m(t)dt\right] sin\omega_c t] \tag{1.61}$$

$$x_{NBPM} \approx A[cos\omega_c t - k_p m(t) sin\omega_c t] \tag{1.62}$$

Both equations are similar to the expression of AM wave. It is possible to generate NBFM and NBPM by using DSB-SC modulator.

Further, for comparison, phasor diagrams of NBFM and AM are depicted in Fig. 1.31, respectively, for sinusoidal modulation. The carrier phasor is used as reference to plot the same. For NBFM, it can be observed that the two side-band phasors add to be at right angle to the carrier phasor leading to a resultant phasor that is out of phase to the carrier but approximately of the same amplitude. On the other hand, the resultant phasor in AM is of different amplitude compared to carrier but is collinear to it. A comparison of NBFM and AM is also provided in table below.

Parameter	NBFM	AM
Bandwidth	$2\,f_m$	$2\,f_m$
Carrier Power	$\frac{A_c}{2}$	$\frac{A_c}{2}$
Power in Sidebands	$\frac{A_c\,\mu}{4}$	$\frac{A_c\,\beta}{4}$
Phase of Resultant Phasors	Right angle to carrier phasor	In phase with carrier phasor
Amplitude of Resultant Phasors	Approximately of same amplitude as carrier phasor	Different amplitude compared to carrier phasor

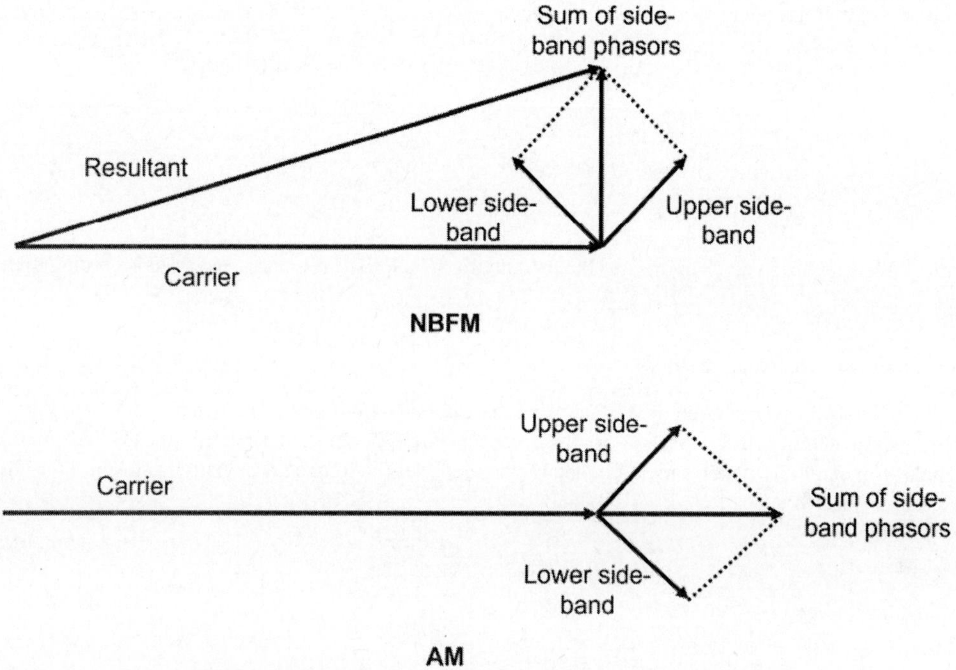

Fig. 1.31 Phasor comparison of NBFM and AM

1.11.2 Wideband angel modulated signal

The wideband FM generation are out into two categories as

i. Direct method
ii. Indirect method

Direct method

In the direct method of generating FM signal, the modulating signal directly control the carrier frequency. The FM signal are generating using a voltage control oscillator (VCO) in direct method. In a VCO, the frequency is controlled by the external voltage.

The FM signal can be generated using the modulating signal $m(t)$ as a input external signal.

$$\omega_i = \omega_c + k_f m(t)$$

Another way of generating FM signal is to vary the reactance (C and L) of the resonant circuit of an oscillator. The oscillator circuit use the parallel tuned L-C circuit and the frequency of the carrier generation is governed by the

$$\omega_c = \frac{1}{\sqrt{LC}}$$

The carrier frequency can vary according to the modulating signal $m(t)$ if L and C is varied according to $m(t)$.

Indirect method (Armstrong method)

In this method, NBFM is converted to WBFM using the additional multiplier. The Armstrong method provides a high frequency stability since crystal oscillator is used for the generation of carrier frequency. The multiplier circuit apart from multiplying the carrier frequency also increases the frequency deviation and hence the NBFM is converted to WBFM.

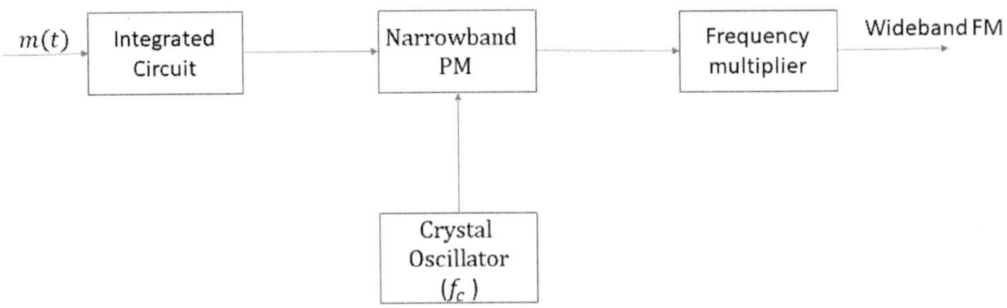

Fig. 1.32 FM generation using Armstrong

EXERCISES

Multiple Choice Questions

1.1 The Dirac delta function $\delta(t)$ is defined as

a) $\delta(t) = \begin{cases} 1, & t = 0 \\ 0, & \text{otherwise} \end{cases}$

b) (b) $\delta(t) = \begin{cases} \infty, & t = 0 \\ 0, & \text{otherwise} \end{cases}$

c) $\Delta(t) = \begin{cases} 1, & t = 0 \\ 0, & \text{otherwise} \end{cases}$ and $\int_{-\infty}^{\infty} \delta(t)\, dt = 1$

d) $\delta(t) = \begin{cases} \infty, & t = 0 \\ 0, & \text{otherwise} \end{cases}$ and $\int_{-\infty}^{\infty} \delta(t)\, dt = 1$

Ans=(d)

1.12 Which one of the following is the correct statement for the given system equation $y(t) = ax(t) + b$?

a) Linear for any value of b
b) Linear for b > 0
c) Non-linear
d) Linear if b < 0

Ans(c)

1.13 Which of the following systems are time invariant with input x(t) and output y(t)

a) $y(t) = tx(t)$
b) $y(t) = x(-t)$
c) $y(t) = x(t/2)$
d) $y(t) = ax(t)$

Ans (d)

1.14 Convolution of x(t+5) with impulse function $\delta(t-8)$ is equal to

a) $x(t-13)$
b) $x(t+13)$
c) $x(t-3)$
d) $x(t+3)$
e) $x(t+13)$

1.15 The modulating frequency in frequency modulation is increased from 10 kHz to 20 kHz. The bandwidth is

a) doubled
b) halved

c) **increased by 20 kHz**
d) increased tremendously

1.16 In phasor representation of FM, the angle of carrier phasor is varied while in AM the length of the carrier phasor is varied.

a) **True**
b) False

Short and Long Answer Type Questions

1.1 If the energy of the signal $x(t)$ is E then calculate the energy of the signal $x(-at + b)$.

1.2 A triangular pulse is given in figure below.

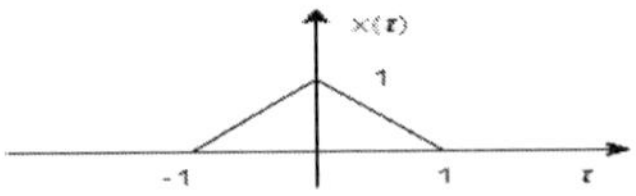

Sketch the signal $x(-2t + 1)$.

1.3 Derive Fourier transform expression for $x(t)=e^{at}u(t)$.

1.4 Explain the conditions for existence of Fourier Transform.

1.5 In a tone-modulated angle modulation, the modulated signal $x_c(t)$ is

$$x_c(t) = A\cos(2\pi f_c t + \beta \sin(2\pi f_m t))$$

when $\beta \ll 1$, we have narrow band angle modulation. Find the spectrum of the narrow band angle modulated signal. Also, draw the phasor representation.

1.6 A carrier is angle modulated by sum of two sinusoidal signals

$$x_c(t) = A\cos(2\pi f_c t + \beta_1 \sin(2\pi f_{m1} t) + \beta_2 \sin(2\pi f_{m2} t))$$

Where the frequencies of two sinusoids are not harmonically related. Find the spectrum of $x_c(t)$.

1.7 Find the normalized average power in angle-modulated signal with sinusoidal modulation.

PRACTICAL

Experiment: -1 Amplitude modulation using DSB TC modulator

Objective

On the completion of this unit, you will be able to study amplitude modulation using DSB modulator.

Discussion on Fundamentals

This circuit investigates how amplitude modulation is performed using double side band modulation. This circuit modulates the frequency of a carrier sine wave, according to the amplitude of audio signal applied to its audio input.

This circuit uses IC 1496, which acts as a frequency multiplier. In DSB TC, carrier also gets transmitted with the information. DSB TC consists of carrier, upper side band & lower side band. Multiplier can act as balanced (suppressed carrier) or standard AM (transmitted carrier). This is achieved by switch closed (TC) & open (SC)

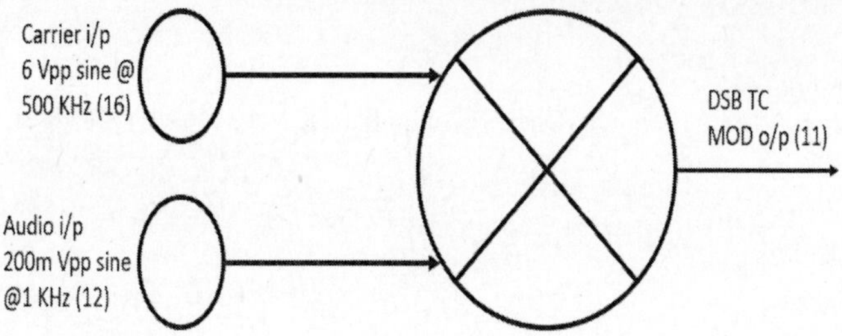

Fig. i Block diagram for DSB TC modulator

Procedure

WW **Wiring sequence (mod)**: - +12 V-1, -12 V-3, GND-2, Audio O/P-12, FG o/p-16, 11-CRO (+)-11, CRO (GND)-32

1 Connect +12 V, −12 V&GND from master unit to banana socket I, 3 & 2 respectively.

2 Make connections as per wiring seq. & set up as shown in Fig (i).

3 Ensure that all switch faults are in OFF position.

4 Keep TC/SC switch i.e. swl @ TC position.

5 Apply audio input (sine wave@1 @Hz;200mVpp) to socket 12 from DTFFG III (audio frequency generator) from master unit.

6 Apply carrier input (500KHz, sin 6Vpp-keep time/ 1 div. = 2u/s & set 1 div.) to socket 16 from master unit function generator (DTFFG III).

7 Observe demodulated o/p at socket 31.

Fig. ii DSB TC modulator waveform

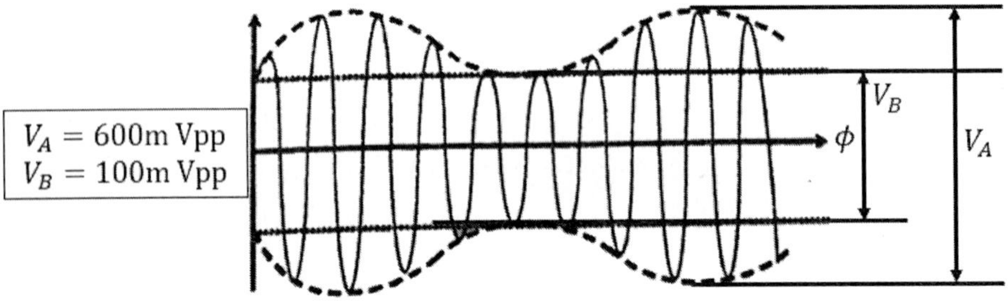

8 Vary applied audio around 1KHz & confirm its being received at demodulated o/p.

Note:
1) DC shift may be observed at socket no.31 due to rectifier followed by LPF action. Observe at AC coupled on CRO to remove DC shift.

2) Give F.G. input as specified in procedure & not exceed audio signal amplitude more than 0.2 Vpp else you observe distortion in demodulation waveform.

Troubleshooting: -
If you are not getting required DSB-TC output at 11, Vary preset VR1 on PCB very precisely as it is factory set.

Equipment's:

1) CM6 Panel

2) CRO

3) Master Unit

Conclusion: Experiment shows how DSB TC gets demodulated using simple diode detector.

Experiment: - 2 DSB TC amplitude demodulation

Objective

On the completion of this unit you will be able to study amplitude demodulation of DSB TC modulator using envelop detector.

Discussion of Fundamentals

1) Envelope detection is usually used in MW broadcast receivers. This circuit investigates how amplitude demodulation is performed using simple diode detector circuit. This requires correct choice of RC time constant. If RC is too large (long time coastant) the output cannot follow rapid decreases in the waveform. If RC is too small (short time constant) it responds too quickly & the carrier leaks into the output giving a large amplitude ripple.

2) since we have used silicon diode, which needs 0.TV to conduct, any incoming wave at socket 30 will have to be amplified using amplifier hence an inverting op-amp is used at socket 30 before applying its output to diode anode(IN4148)

Fig. i Block

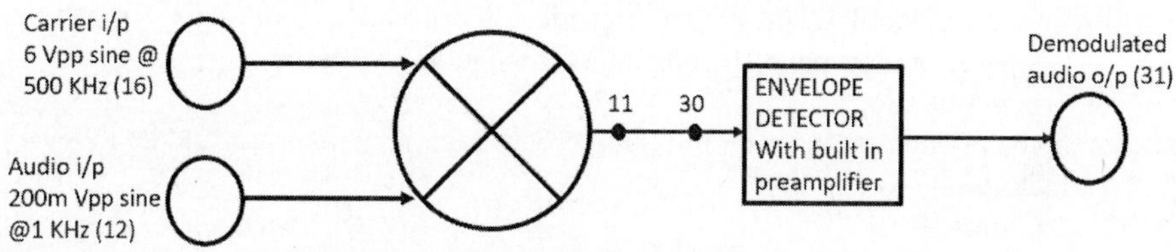

diagram for DSB TC demodulator

Procedure:

WW **Wiring sequence (mod)**: - +12 V-1, -12 V-3, GND-2, Audio O/P-12, FG o/p-16, 11-CRO (+), CRO (GND)-32

WW **Wiring sequence (demod)**: - Do not disconnect above wiring (except CRO), 11-30, CRO (+)-31, CRO (GND)-32

1 Connect +12 V, −12 V&GND from master unit to banana socket I, 3 & 2 rospectively.

2 Make connections as per wiring seq. & set up as shown in fig1.
3). Ensure that all switch faults are in OFF position.

3 Keep TC/SC switch i.e.sw1 @ TC position.

4 Apply audio input (sinewave@1 @Hz; 200mVpp) to socket 12 from DTFFG III (audio frequency generator) from master unit.

5 Apply carrier input (500KHz, sin 6Vpp − keep time/ 1 div. = 2u/s & set 1 div.) to socket 16 from master unit function generator (DTFFG III).

6 Observe demodulated o/p at socket 31.

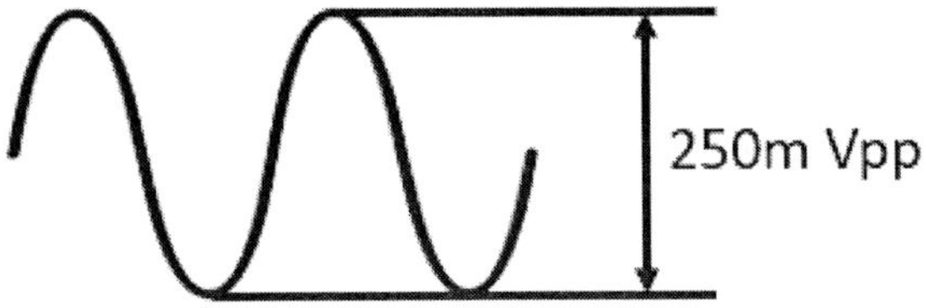

Fig. ii Audio output for DSB TC demodulator

7 Vary applied audio around 1KHz & confirm its being received at demodulated o/p.

Note:

1)DC shift may be observe at socket no.31 due to rectifier followed by LPF action. Observe at AC coupled on CRO to remove DC shift.

2)Give F.G. input as specified in procedure & not exceed audio signal amplitude more than 0.2 Vpp else you observe distortion in demodulation waveform.

Troubleshooting:-

If you are not getting required DSB-TC output at 11, Vary preset VRI on PCB very precisely as it is factory set.

Equipment's:

1) CM6 Panel
2) CRO
3) XPO-COM Master Unit

Conclusion:

Experiment shows how DSB TC gets demodulated using simple diode detector.

Experiment: -3 Amplitude modulation using DSB SC modulator

Objective

On the completion of this unit you will be able to study amplitude modulation with DSB SC modulator.

Discussion of Fundamentals

This circuit investigates how amplitude modulation is performed DSB SC modulator circuit. The main disadvantage of DSB TC is more bandwidth is required for transmission, which is not useful. Hence DSB SC is used. In DSB SC carrier gets suppressed, only two side bands are present.

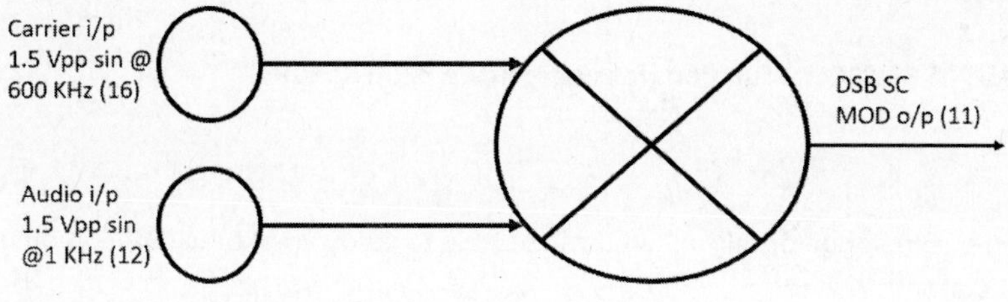

Fig. i Block diagram for DSB SC modulator

Procedure

1 Connect +12 V, −12 V& GND from master unit to banana Socket 1,3 & 2 respectively.

2 Ensure that all switch faults are in OFF position.

3 Apply audio input 1KHz, 1.5Vpp sine from audio generator to socket 12 from DTFFG III.

4 Apply carrier input 600KHz, (keep Time/div. = 0.5 us and set 3.3 div.) 1.5 V pp sin from DTFFG III function generator to socket 16.

5 Keep switch SW1 at SC position.

6 Connect CRO between socket 11 & 32 i.e. GND & observe modulated output (DSB SC).

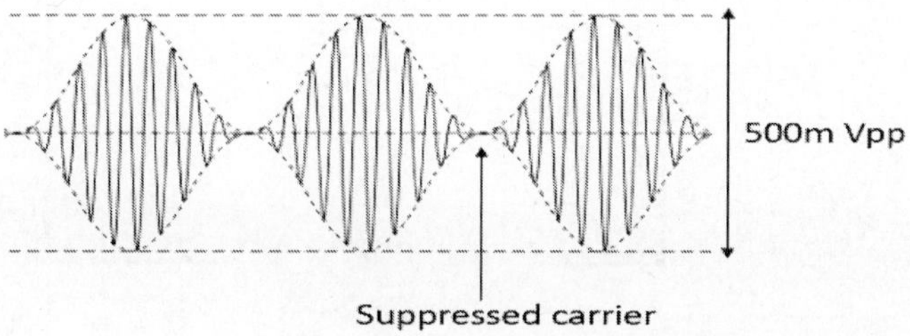

Fig. ii DSB SC Modulator waveform

Note:

When modulating input = zero (@ zero crossing) then carrier output is zero.

Equipment's:

1) CM6 Panel

2) CRO

3) XPO-COM Master Unit

Conclusion:

Experiment shows that how carrier gets suppressed inactivate DSB SC modulator.

Experiment: - 4 SSB SC modulation (for upper/lower side band)

Objective

On the completion of this unit you will be able to study amplitude modulation using SSB modulator.

Discussion of Fundamentals

Single sideband suppressed carrier modulation was the basis for along distance telephone communications up until last decade. Before arrival of digital telephony, the sideband supported hundreds & thousands of individual telephone conversations.

This section investigates how amplitude modulation is performed using SSB modulator circuit. In this circuit outputs of two DSB modulators are quadrature (90 degree) added to get SSB SC o p.

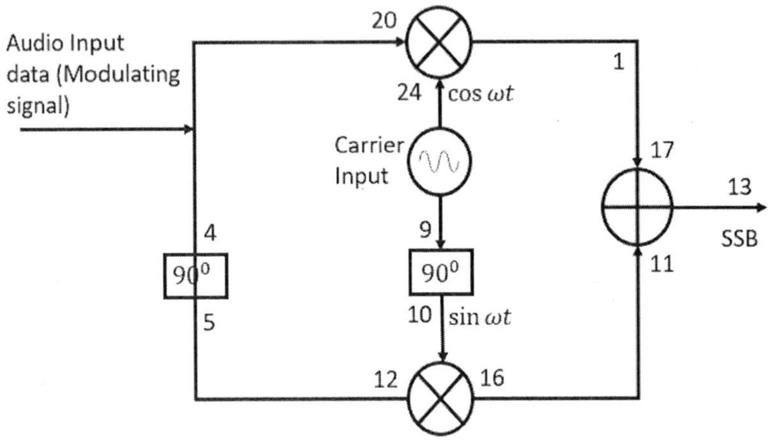

subtraction is achieved by first phase shifting carrier & audio by 90 degree. Obviously this function has to be mathematically accurate to delete one of the two bands (upper & lower) Any error will cause o/p (SSB) to have some contamination (looks like AM or DSB TC) By experiment after observing the frequency limitation of multiplier 1C, the carrier is limited to 600KHz. Use the preset knob of both phase shifter carefully & softly to achieve best results for SSB. Also amplitude of both carrier & audio as well as 90 degree phase shifter counter parts of there should be identical else contamination occurs. In fact inability to set up two identical multiplier blocks vis-à-vis amplitude & phase is the source of error.

Figure shows block diagram for SSB SC.

Fig. i Block Diagram for SSB SC modulator

Procedure:

WW **Wiring sequence (mod)**: - +12 V-1, -12 V-3, GND-2, Audio O/P-4 and 20, FG o/p-9 and 24, 10-16,5-12, 17-18, CRO (+)-13-FC(+), CRO (GND)-32, FC(-)28

1 Connect +12 V, −12 V&GND from master unit to banana socket 1,3 & 2 resply.

2 Make connections as per wiring sequence & set up as shown in figl.

3 Ensure that all switch faults are in OFF position.

4 Keep TC/SC switches i.e. sw1 & sw2 (a) SC position.

5 Keep SW 4 switch at up position.

6 Apply audio input (sin wave (a) 1KHz, 1.5Vpp) to socket 4 & 20 i.e. audio o/p from DTFFG III (audio frequency generator)

7 Apply carrier input (600KHz, sin 1.5Vpp – keep time/div = 0.5 us and set 3.3 div.) to socket 9,24 from master unit function generator (DTFFG III).

8 Observe SSB SC output at socket 13 on CRO as well as counter. (Adjust P1 & VR6 so thatyou will get better SSB o/p without any contamination which looks like DSB otherwise) Fmod-Fcarrier-Faudio (SSB down) You may vary audio signal by ±500 Hz & observe corresponding change in reading of FC.

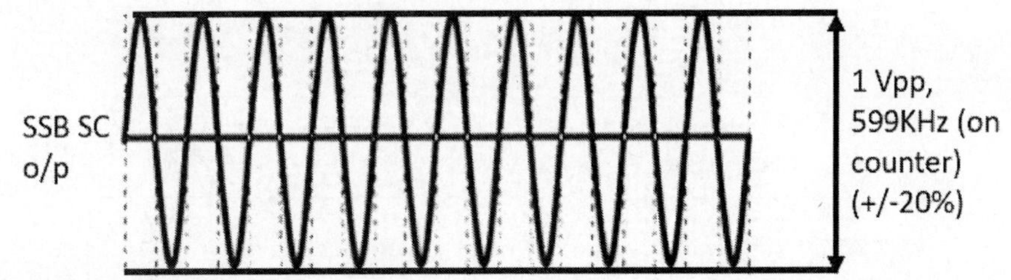

Fig. ii Block diagram SSB SC modulator waveform

Equipment's:

1) CM6 Panel
2) CRO
3) XPO-COM Master Unit

Conclusion:

Experiment shows how SSB SC modulation is derived using two DSB SC modulators. The frequency counter shows one of two sidebands only even if there is slight contamination from other sideband. The frequency counter showing carrier (600KHz) !ess by 1 KHz (mod sin) is the correct test of SSB even though due to imperfection of implementing OPAMPS slight amplitude variation will be seen. You can use frequency measurement facility available on any channel of DSO. For this purpose, use any lab tabletop counter (FC50M)

Experiment: - 5 DSB SC demodulation

Objective

On the completion of this unit, you will be able to study amplitude demodulation of DSB SC modulator using product detector.

Discussion of Fundamentals

The audio signal can be recovered with a further multiplication by a carrier.It is essential for the frequency and phase of this carrier to be exactly the same as (synchronous with) the original, otherwise there will be frequency and phase distortion of the signal. This circuit investigates how amplitude demodulation of DSB SC is performed using frequency multiplier circuit with built in RC(LPF)

Fig. i

Block

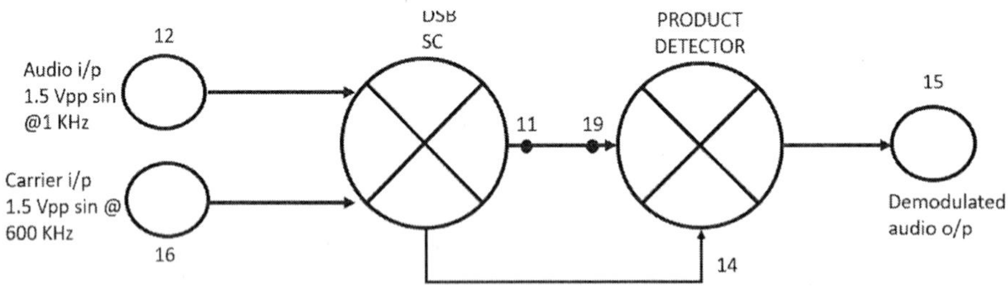

diagram of DSB SC modulator

Procedure:

WW **Wiring sequence (mod)**: - +12 V-1, -12 V-3, GND-2, Audio O/P-12, FG o/p-16, 11-CRO (+), CRO (GND)-32

WW **Wiring sequence (demod)**: - Do not disconnect above wiring, 16-14, 11-19, CRO (+)-15, CRO (GND)-32

1 Connect $+12$ V, -12 V & GND from master unit to banana socket 1,3 & 2 respectively.

2 Make connections as per wiring sequence & set up as shown in fig.

3 Ensure that all switch faults are in OFF position.

4 Keep TC/SC switches i.e.sw1&sw2@SC position.

5 Apply audio input (sin wave@1 1KHz, 1.5Vpp) to socket 12 from DTFFG III (audio frequency generator)

6 Apply carrier input (600KHz, sin 1.5Vpp – keep time/div. = 0.5u/s & set 3.3div.) to socket 16 from master unit function generator (DTFFG III).

7 Observe demodulated o/p at socket 15.
Fig 6.6.2 Audio output for DSB SC mod-demod

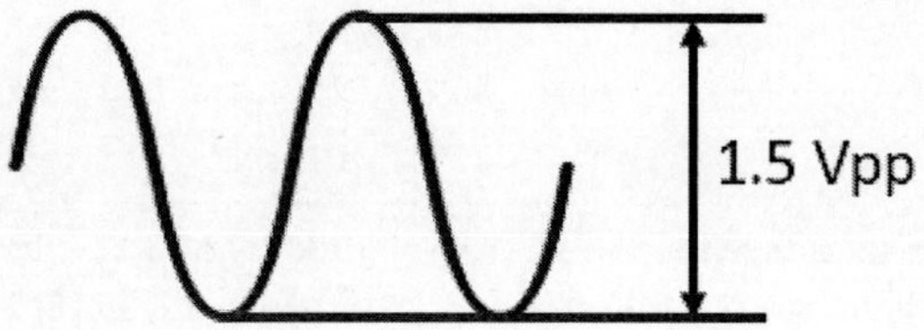

Fig. ii Audio output of DSB SC demodulator

Note:-

If demodulated output at 15 is not in proper shape, adjust carrier signal amplitude upto 1Vpp to observe clear sine wave.

Equipments:

Equipment's:

1) CM6 Panel
2) CRO
3) XPO-COM Master Unit

Conclusion:

Experiment shows how DSB SC gets demodulated using product detector.

Experiment: - 6 SSB SC demodulation

Objective

On the completion of this unit you will be able to study amplitude demodulation of SSB SC modulator using product detector.

Discussion of Fundamentals

This circuit investigates how amplitude demodulation of SSB SC is performed using frequency multiplier circuit.

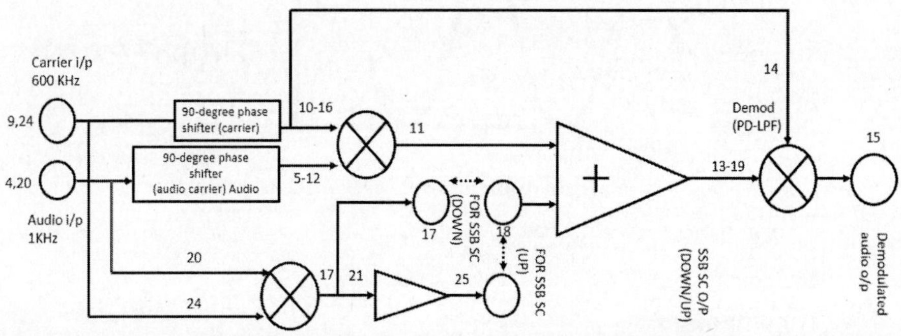

Fig. i Block diagram of SSB SC demodulator

Procedure:

WW **Wiring sequence for SSB SC DOWN (mod)**: - $+12\text{ V}-1, -12\text{ V}-3$, GND-2, Audio o/p-4 & 20, FG o/p -9 & 24, 10-16, 5-12, 1718, 13 -CRO (+), CRO

WW **Wiring sequence for SSB SC UP (mod)**: - $+12\text{ V}-1, -12\text{ V}-3, \text{GND}-$ 2, Audio o/p-4 & 20, FG o/p 9 & 24, 10-16, 5-12,17-21, 25-18, CRO (+) -13-FC (+), CRO (GND)-32, FC (-)-28

WW **Wiring sequence (demod)**: - Do not disconnect wiring, 13-19. 16*14, CRO (4)-15, CRO (GND)-32

Procedure :

1. Connect $+12$ V, -12 V & GND from master unit to banana socket 1,3 & 2 resp.
2. Make connections as per wiring sequence & set up as shown in figI.
3. Ensure that all switch faults are in OFF position.
4. Keep TC/SC switches i.e.sw 1 & sw2 (a) SC position.
5. Keep SW4 (phase control for audio/carrier) in upward direction.
6. Apply audio input (sine wave@1KHz, 800mVp p) to socket 4, 19 from DTFFG III (audio frequency generator).
7. Apply carrier input (600KHz, sin 1.5Vpp-keep time/div $= 0.5$ us & set 3.3 div.) to socket 9 & 24 from master unit function generator (DTFFG III).
8. Observe audio $0/\text{p}$ at socket 15 .

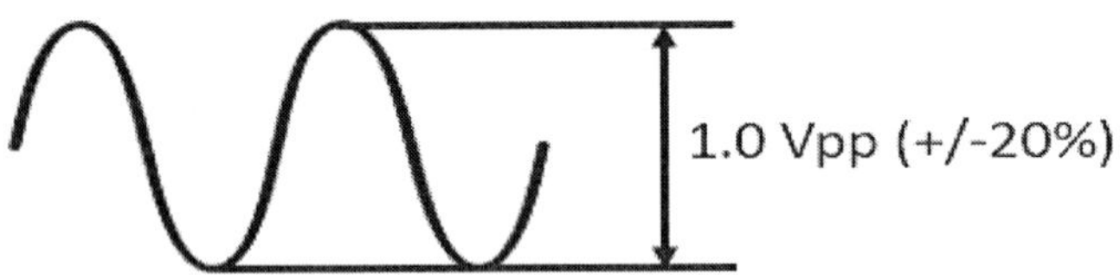

Fig. ii Audio output for SSB-SC demodulator

Equipment's:

1) CM6 Panel

2) CRO

3) Master Unit

Conclusion:
Experiment shows how SSB SC gets demodulated using product detector.

Experiment: - 7 Frequency modulation & demodulation using Reactance modulator& Detuned resonant detector

Objective
On the completion of this unit you will be able to study frequency modulation & demodulation using Reactance modulator & detuned resonant detector.

Discussion of Fundamentals
Detuned Resonant Circuit Detector
This is the most basic type of demodulator. A parallel tuned circuit is deliberately detuned so that the incoming carrier occurs approximately halfway up the slope of the response's left-hand side.

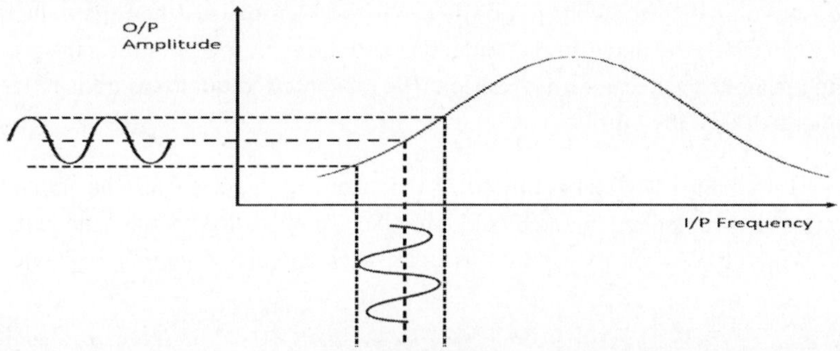

Fig. i Output amplitude vs frequency

As shown in Fig. *i* , as the input frequency changes, the amplitude of the output signal will increase and decrease. For instance, if the frequency of the incoming signal increased, the operating point on the diagram would shift to the right. This would result in an increase in the output signal's amplitude. The output of an FM signal will therefore be an amplitude modulated signal.

Fig ii Detuned Resonant Circuit

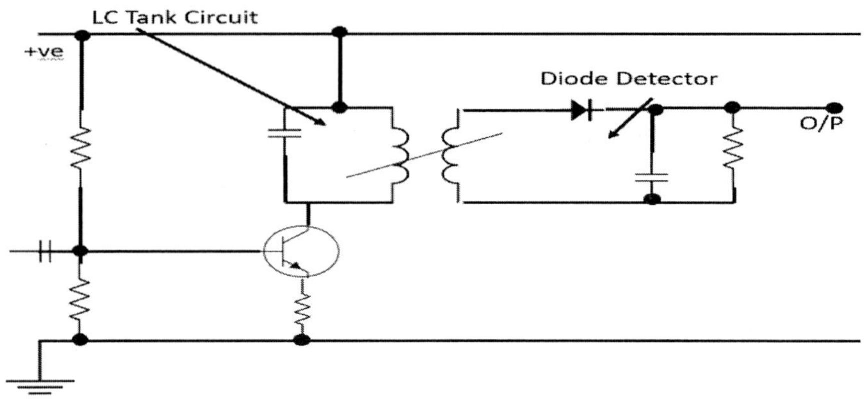

Above Fig. ii shows a diagram of the Detuned Resonant Circuit Detector's circuitry. When broken down, the procedure becomes very clear. The FM input is applied to the base of the transistor, and the detuned resonant circuit is located in the collector. It also includes the loading effect caused by the other transformer secondary winding. The signal at the collector of the transistor is passed to the diode detector with an amplitude modulated component. In the diagram, the diode conducts whenever the input signal at its anode is greater than the voltage on the capacitor's top plate. When the voltage falls below the voltage of the capacitor, the diode stops conducting and the voltage across the capacitor leaks away until the next time the input signal can turn it back on. The unwanted frequency is transmitted to the Low Pass Filter/Amplifier block via the output.

One drawback is that any noise spikes present in the incoming signal will also be amplified and appear at the output after being amplified by the diode detector. The AM noise must be eliminated before the demodulator's input if we want to avoid this issue. Using an amplifier limiter circuit, we accomplish this.

Note:

In many communications experiment, the output of desired experiment is DC shifted so DC blocking is required. If demodulated output is of low amplitude level, it need to connect demodulator output to HPF AC amplifier (1.6K) of NGLPF) & using preset adjust amplitude level.

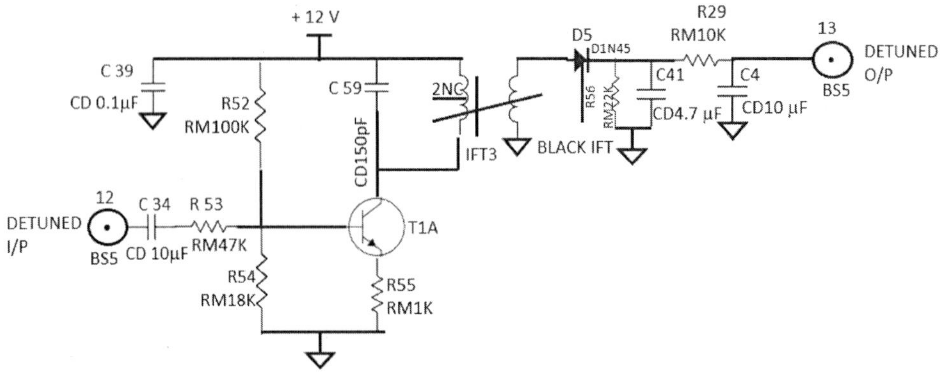

Fig. iii Circuit diagram for detuned resonant detector

Procedure:

WW **Wiring sequence (mod.):** - $+12\ V - 1, -12\ V - 3, GND - 2$, Apply sine wave (a) $1KH, 1Vpp$) to socket $14{,}11 - CRO(+), CRO(GND) - 21$

WW **Wiring sequence (Demod):** - (Keep mod. Connections as it is.) $11 -$ $4{,}6 - 22{,}23 - 12{,}13 -$ LP1 (on NGLPF panel on MU), CRO (+)-2P|, $CRO(GND) - 21FGo/p - 14.$
Wiring sequence for Stand-alone assembly: 11-4, 6-22, 23-12, 13-27, CRO $(+) - 28$, CRO (GND)-21, FG o/p-14.

1 Connect $+12\ V, -12\ V$ & GND from master unit to banana socket 1,3&2 resply (No need in case of stand-alone).

2 Make connections as per wiring sequence & set up as shown in figl.

3 Ensure that all switch faults are in OFF position.

4 Keep DC bias preset at approx. middle position.

5 Apply audio input (sine wave @1KHz, 1Vpp) to socket 14 i.e. audio input.

6 Connect CRO between socket 2P1, NGLPF on MU (or, at socket 27 in case of Stand-alone) & GND. Observe demodulated s/g at output of detuned resonant detector.
Table 5.1 Observation Table for Detuned Resonant Detector

7 Now fill table 5.2. & Note down readings for different frequencies (by keeping audio i/p voltage constant = $0.65\ V$). Plot graph of Vo/pVs Freq. note down readings for different input voltages (by keeping audio input frequency constant $= 1KHz$) & plot graph of $Vo/pVsI/P$ voltage from table 5.3
Table 5.2 Audio input (i/p) (Hz) Vs output (o/p) Table 5.3 Audio input (i/p) (volts) Vs output (o/p)

Audio isp (Hz)	Detuned (Vpp)		
100		Audio i/p (Vpp)	Detuned (Vpp)
1		0.1	
1 K		0.2	

2 K		1	
3 K		1	
4 K		1	
5 K		1.0	

Equipment's:

1) CM6 Panel
2) CRO
3) XPO-COM Master Unit

Conclusion:

Experiment shows how modulated signal gets demodulated using reactance modulator & detuned resonant detector.

KNOW MORE

Fundamentally, communication is known as exchange of information to one person to another through verbal or non-verbal means. The need for communication arises in the form of listening, writing, speaking, and reading. The evolution communication started with communicating through fire, then pigeons, in ancient times. Another form of communication was during the era of camel messengers, where messages were delivered to one person to another through camel riding.

Martin Cooper

In the 19th century, the first electrical based communication started which was through telegraph and this removed the barrier of geography. Also, Martin Cooper brings the revolution in the telephone industry by inventing the very first handheld cellular mobile phone.

Timelines

1830-1832:	Telegraph's invention by Joseph Henry and Pavel schilling
1876:	Telephones were invented by Alexander G. Bell
1887:	Heinrich Hertz detected electromagnetics.
1901:	Marconi made the first radio telegraph communication over the Atlantic.
1906:	Transmission using amplitude modulation for the first time by Reginald A. Fessenden

1928: Philo Farnsworth demonstrated first television broadcasting

1933: FM broadcasting was invented by Edwin Armstrong.

1969: The U.S. government created an early form of the Internet called ARPANET.

Application (Real life/Industrial)

Communication systems that employ a continuous signal to transmit voice, data, image, signal, or video information frequently use analogue signals. Analog transmission can be divided into two categories that are both based on how data is adjusted to combine an input signal with a carrier signal. Amplitude modulation and frequency modulation are the two methods. Analog circuits use these techniques to interact with the outside world and to precisely record and process these signals in electronics, just how the human body uses its eyes and ears to collect sensory information.

References and Suggested Reading

1. B. P. Lathi, and Zhi Ding, "Modern Digital and Analog Communication Systems", Oxford University Press, 2010.
2. Simon Haykin, "Communication Systems", Wiley, 2007.
3. Herbert Taub, Donald L. Schilling, and Gautam Saha, "Principal of Communication Systems", McGraw Hill Education, 2017.
4. P.F. Panter, "Modulation, Noise and Spectral Analysis", McGraw-Hill, New York, 1965.
5. N Abramson, "Bandwidth and Spectra Phase- and Frequency-Modulated Waves",IEEE Transaction Communication System, CS_11, PP. 407-414, 1963.
6. https://nptel.ac.in/courses/117102059
7. https://ocw.mit.edu/courses/16-36-communication-systems-engineering-spring-2009/pages/lecture-notes.

2 NOISE

UNIT SPECIFICS

Through this unit we have discussed the following aspects:

- *Review of probability and random process,*
- *Gaussian and white noise characteristics,*
- *Noise in amplitude modulation systems,*
- *Noise in frequency modulation systems,*
- *Pre-emphasis and De-emphasis,*
- *Threshold effect in angle modulation.*

The practical applications of the topics are discussed for generating further curiosity and creativity as well as improving problem solving capacity.

Besides giving a large number of multiple-choice questions as well as questions of short and long answer types marked in two categories following lower and higher order of Bloom's taxonomy, assignments through a number of numerical problems, a list of references and suggested readings are given in the unit so that one can go through them for practice. It is important to note that for getting more information on various topics of interest some QR codes have been provided in different sections which can be scanned for relevant supportive knowledge.

After the related practical, based on the content, there is a "Know More" section. This section has been carefully designed so that the supplementary information provided in this part becomes beneficial for the users of the book. This section mainly highlights the initial activity, examples of some interesting facts, analogy, history of the development of the subject focusing the salient observations and finding, timelines starting from the development of the concerned topics up to the recent time, applications of the subject matter for our day-to-day real life or/and industrial applications on variety of aspects, case study related to environmental, sustainability, social and ethical issues whichever applicable, and finally inquisitiveness and curiosity topics of the unit.

RATIONALE

This unit on Noise helps students to get a primary idea about the effect of noise in communication systems including the one in different modulation schemes and the methods to reduce those effects. We have started with basics and pre-requisites required for the topic. First, probability and random process is thoroughly revised before moving on to the more complex topics. Then we have covered the characteristics of the most widely considered noise in communications system i.e. Gaussian and white noise. Then we have analysed the effect of noise in amplitude modulation systems and frequency modulation systems. To mitigate the effects of noise we have covered methods such as pre-emphasis and de-emphasis. At last, we have covered threshold effect in angle modulation

In communication system the distortion and impact of noise is unavoidable, but one can explore methods to minimize or mitigate those effects. Throughout this chapter, we attempt to learn about the characteristics of noise, its impact on different modulation systems and how we can minimize those impacts for optimum performance.

PRE-REQUISITES

Mathematics: Calculus (Class XII)

Physics: Basic of Communication (Class XII)

UNIT OUTCOMES

List of outcomes of this unit is as follows:

U1-O1: Describe basics of probability and random Process.

U1-O2: Explain the additive white Gaussian noise

U1-O3: Describe noise in amplitude modulation

U1-O4: Explain noise in frequency modulation

U1-O5: Explain effects of noise in communication system

This table below need to be filled after discussion as marked with red.

Unit-1 Outcomes	EXPECTED MAPPING WITH COURSE OUTCOMES (1- Weak Correlation; 2- Medium correlation; 3- Strong Correlation)					
	CO-1	CO-2	CO-3	CO-4	CO-5	CO-6
U1-O1	3	3	3	-	3	1
U1-O2	1	1	2	2	1	-
U1-O3	2	1	3	1	2	1
U1-O4	-	-	3	1	2	2
U1-O5	3	3	3	-	3	1

2.1 PROBABILITY

Probability can be defined as the likelihood of the occurrence of an event. Pr(A), which represents the probability of an event A, has the following definition:

$$\Pr(A) = \frac{\text{Number of favorable outcomes}}{\text{Total number of outcomes}} \tag{2.1}$$

2.1.1 Properties of Probability

Let the set of outcomes is denoted by a letter (e.g., A), and its probability is represented by Pr(A). Then the rules for probabilities are:

Property 1: - $Pr(\Omega) = 1$; the sum of all probabilities is 1

Property 2: - $Pr(A) \geq 0$; probability is always positive.

Property 3: - If two sets of outcomes are disjoint (mutually exclusive) then $Pr(A_i \cup A_j) = Pr(A_i) + Pr(A_j)$

The probability of the union of the sets is the sum of the individual probabilities. In other words, the chance of having either one is equal to the total of the probabilities for each if having A excludes having B and vice versa (think of throwing a die, which has six disjoint outcomes).

2.1.2 Conditional Probability

Conditional probability refers to the situation in which the likelihood that one event will occur depends on the likelihood that another event will occur.

Given that event B has already happened, the conditional probability of event A is given by,

$$Pr(A|B) = Pr(A \cap B) / Pr(B) \tag{2.2}$$

Let us consider an example. We define three possible events:

E: A earthquake has occurred.

G: A foreshock has occurred.

D: A large/distinctive earthquake will occur.

We would categorise a small background shock that happened by coincidence right before the distinctive earthquake as a foreshock. As a result, E and D cannot coexist because they are independent. The same is also true for E and G. The conditional probability of D, given either G or E, is the probability we are looking for (because we do not know which has occurred).

We can write the probability $Pr(D|G \cup E)$ as the probability of occurrence of a large earthquake given a background earthquake and/or a foreshock has already occurred. This can be further expanded as,

$$Pr(D|G \cup E) = Pr(D \cap (G \cup E))\ Pr(G \cup E)$$

As F and E are disjoint, the probability of their union is the sum of the individual probabilities (from property 3), allowing us to write the numerator as $Pr((D \cap G) \cup (D \cap E)) = Pr(D \cap G) + Pr(D \cap E) = Pr(D \cap G)$ where the term $Pr(D \cap E)$ is eliminated as E and D are disjoint.

From the definition of conditional probability, $Pr(D \cap G) = Pr(G|D)P(D)$ where $Pr(G|D)$ is the probability that a mainshock is preceded by a foreshock. As G and E are disjoint, we can write the denominator as

$$Pr(G \cup E) = Pr(G) + Pr(E)$$

As a foreshock cannot, by definition, occur without a mainshock, the intersection of D and G is F, and therefore

$$Pr(G \cap D) = Pr(G|D)Pr(D) = Pr(G)$$

Finally, we can get,

$$Pr(D|G \cup E) = Pr(G)\ Pr(G) + Pr(E) = Pr(D)Pr(G|D)\ Pr(G|D)Pr(D) + Pr(E)$$

2.1.3 Probability of Statistically Independent Event

If the likelihood of one event does not influence the likelihood of another event occurring, then X and Y are two statistically independent events

$$Pr(Y|X) = Pr(XY) \backslash Pr(X).$$

Now since the occurrence of Y doesn't depend on the occurrence of X therefore $Pr(XY)=Pr(X)*Pr(Y)$

$$Pr(Y|X) = Pr(Y)$$

Similarly,

$$Pr(X|Y) = Pr(X)$$

2.2 RANDOM VARIABLES

Suppose there is a sample space S for a random experiment. We can define a single-valued real function X(r) for such a sample space so that each sample point r in the sample space S is translated into a real number. The term "random variable" refers to this X(r) or simply X. An experiment with a random variable will almost always have a range of possible values.

A probabilistic description of X is required, and this description manifests as a function known as the probability density function of X. The probability density function tells the probability of occurrence of a certain value of random variable.

A cumulative distribution function also exists for all random variables, both discrete and continuous. It is a function that indicates, for any value of x, the likelihood that the random variable X is less than or equal to x. For a discrete random variable, the CDF is found by summing up the probabilities.

There are two types of random variables: -

1) Continuous random variable

2) Discrete random variable

2.2.1 Discrete Random Variable

X is said to be a discrete random variable if there are only a finite number of possible values that it can take. The number of children in a household, the Friday night movie attendance, the number of patients at a doctor's office, and the number of faulty light bulbs in a box of ten are a few examples of discrete random variables.

A discrete random variable's probability distribution is a table of the probabilities connected to each of its potential values. The probability function or probability mass function are other names for it.

Suppose a random variable X may take k different values, with the probability that $X = x_i$ defined to be $\Pr(X = x_i) = p_i$. Then p_i should satisfy the following conditions:

1) $1: 0 \leq p_i \leq 1$ for each i
2) $2: p_1 + p_2 + ... + p_k = 1$.

Example: - 1 Let X be a variable that can take the values 1, 2, 3, or 4.
The corresponding probabilities associated with each outcome is given by:

Outcome	1	2	3	4
Probability	0.1	0.3	0.4	0.2

Pr (X = 2 or X = 3) = Pr(X = 2) + Pr(X = 3) = 0.3 + 0.4 = 0.7.
Similarly, the probability that X is greater than 1 = 1 - Pr(X = 1) = 1 - 0.1 = 0.9, by the complement rule.

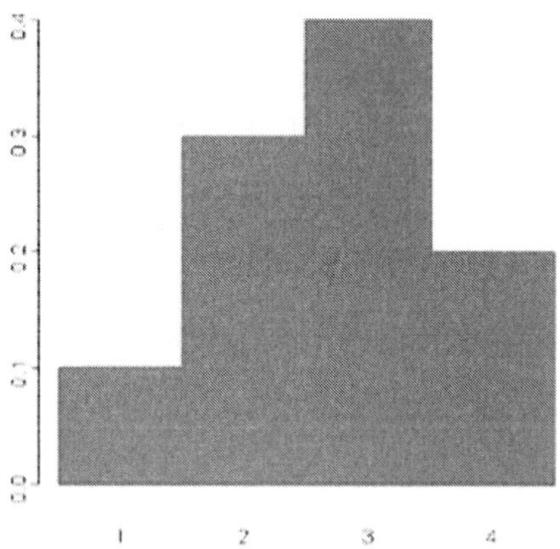

Fig. 2.1 Probability distribution histogram

2.2.2 Continuous Random Variables

If a random variable can take an infinite number of possible values within a certain range then we represent it using a continuous random variable. In most cases, continuous random variables are measurements. Examples include a person's height, weight, orange's sugar content, and mile-run time.

One cannot define a continuous random variable at pinpointed values. Instead, it is represented as area under a curve and specified over a range of values. Since the random variable might take on an endless number of values, the probability of detecting any one particular value is equal to zero.

The following conditions must be met by the curve that represents the function p(x):

1) No negative values exist for the curve ($p(x) > 0$ for all x).
2) The total area under the curve is equal to one.

A density curve is one that meets these requirements.

2.3 STATISTICS OF RANDOM VARIABLE

2.3.1 Mean of Random Variable

The mean of a random variable X is given by μ where μ is defined as,

Mean $(\mu) = \sum X \, Pr(X)$

here X is the set of all possible values the random variable can take and Pr is the corresponding probability for each value of X

2.3.2 Variance of Random Variable

Variance is the spread of X around the mean value (μ),

$$Var(X) = \sigma^2 = E[X^2] - [E(X)]^2$$

where $E[X^2] = \sum X^2 Pr$ and $E[X] = \sum XPr$

and E[X] is the expected value of random variable X.

2.4 RANDOM PROCESS

As is well known, noises cannot be described by deterministic functions of time, and random signals cannot be explicitly described before they occur. However, a random signal or noise may display some regularities over the course of prolonged observation that can be explained in terms of probabilities and statistical averages. Such a case is called random process.

Also, it aids in our understanding of systems involving uncertain signals.

2.4.1 Definition of Random Process

Consider an experiment with sample space S and outcomes λ. If for each outcome λ ∈S, assign a real valued time function X(t, λ), create a random (or stochastic) process. A random process X(t, λ) is function of 2 parameters, time and the outcome λ.

A group of time signals or functions that represent various outcomes of an experiment is a random process.

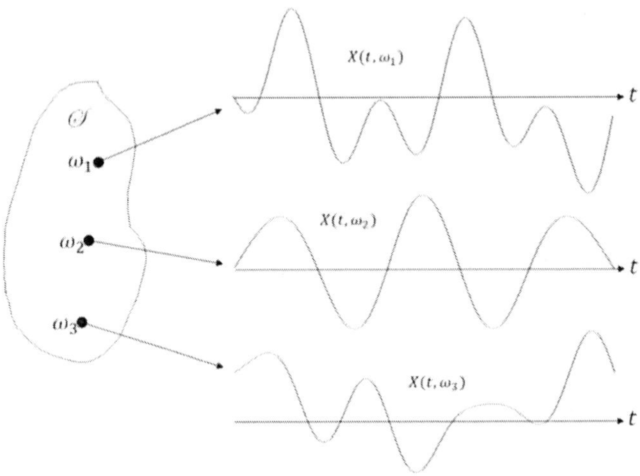

Fig. 2.2 Random process

A random variable which as a function of time is a random process. So, sampling of random process gives random variable.

2.4.2 Stationarity

2.4.2.1 Strict Sense Stationary

If the statistics of any collection of samples do not change over time, a random process X(t) is said to be strictly stationary or stationary in the strict sense. In other words, the joint cdf or pdf of $X(t_1)$, …, $X(t_K)$ is the same as the joint cdf or pdf of $X(t_1+\tau)$,…..,$X(t_K+\tau)$ for any time shift τ, and for all choices of t_1, …,t_K.

The variance and mean of a stationary random process are constants.

Since it is challenging to predict the distribution Jof a random process. Random process is judged by the mean, autocorrelation, and autocovariance functions.

2.4.2.2 Wide Sense Stationary:

If the mean of X(t) is constant and the autocorrelation function only depends on the time difference $\tau = t_2 - t_1$ and not on t_1 and t_2 individually, then random process, X(t) is a wide-sense stationary process (WSS). In a wide-sense stationary process, the choice of the time origin has no bearing on the autocorrelation functions and mean.

Properties Of WSS:

1) $E[X(t1)]=E[X(t2)]$
2) $RX(t1,t2)=RX(t1-t2)$
   where E[.] is the statistical expectation operator
3) Correlations: Assuming that the random processes are WSS.

Autocorrelation Function:

$$R_{xx}(\tau) = E[X(t)X(t+\tau)]$$

Properties of Autocorrelation function:

1) $|R_{xx}(\tau)| \leq R_{xx}(0)$
2) $R_{xx}(\tau) = R_{xx}(-\tau)$
3) $R_{xx}(0) = E[X_2(t)]$

$E[X_2(t)]$ is usually referred to as the *mean-square value.*

Cross-Correlation Function:

$$R_{XY}(\tau) = E[X(t)X(t+\tau)]$$

Properties of Cross-correlation function:

1) $R_{XY}(\tau) = R_{XY}(-\tau)$
2) $|R_{XY}(\tau)| \leq \surd R_{XY}(0).R_{xx}((0)$
3) $|R_{XY}(\tau)| \leq R_{XY}(0).R_{xx}(0) / 2$

Auto Covariance Function:

$$C_{XX}(\tau) = E[\{X(t) - E[X(t)]\}\{X(t+\tau) - E[X(t+\tau)]\}]$$

Cross Covariance Function:

$$C_{XY}(\tau) = E[\{X(t) - E[X(t)]\}\ \{Y(t+\tau) - E[Y(t+\tau)]\}]$$

$= R_{XY}(\tau) - \mu_X \mu_Y$

where μ_X and μ_Y is the mean of the random process X(t) and Y(t).

Two processes X(t) and Y(t) are called (mutually) orthogonal if $R_{XY}(\tau)=0$.

Power Spectral Density:

$$S(\omega) = \lim_{\tau\to\infty} |X(\omega)|^2/\tau$$

Properties of Power Spectral Density:

1) $S_x(-f) = S_x(f)$ for all f, where f is the frequency in Hertz.
2) $S_x(f) \geq 0$, for all f
3) $E[X(t)2] = RX(0) = \int_{-\infty}^{\infty} (f)\, df$
4) $S_x(f)$ is a real function

2.4.3 Additive White Gaussian Noise (AWGN)

Communication system performance degrades as a result of noise. The same is measured by calculating the signal-to-noise ratio (SNR) of the system. The subsequent SNR analysis takes into account the effect of additive noise on the signal at the receiver.

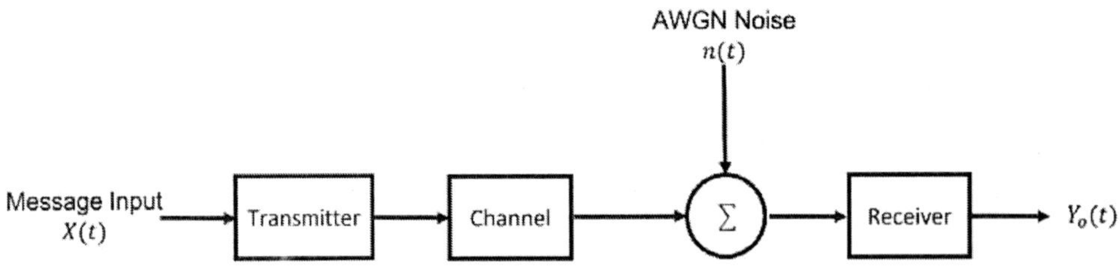

Fig. 2.3 Block diagram of a communication system

Let us assume that a random process $X(t)$ is given as an input message to the transmitter as shown in Fig. (2.3). The receiver is considered to be a linear system which receives the input along with noise introduced by the channel. It is assumed that the noise $n(t)$ is white gaussian, has a zero mean, and has a power spectral density of $\frac{\eta}{2}$ and is uncorrelated with $X(t)$. The output $Y_o(t)$ of the receiver can be expressed as

$$Y_o(t) = X_o(t) + n_o(t) \tag{2.3}$$

where, respectively, $X_o(t)$ and $n_o(t)$ represent the signal and noise components at the receiver's output. The output SNR $\left(\frac{S}{N}\right)_o$ is then defined as

$$\left(\frac{S}{N}\right)_o = \frac{S_o}{N_o} \tag{2.4}$$

where $S_o = E[X_o^2(t)]$ and $N_o = E[n_o^2(t)]$ are the average signal and noise power values at the receiver's output.

2.4.4 Baseband Communication Systems

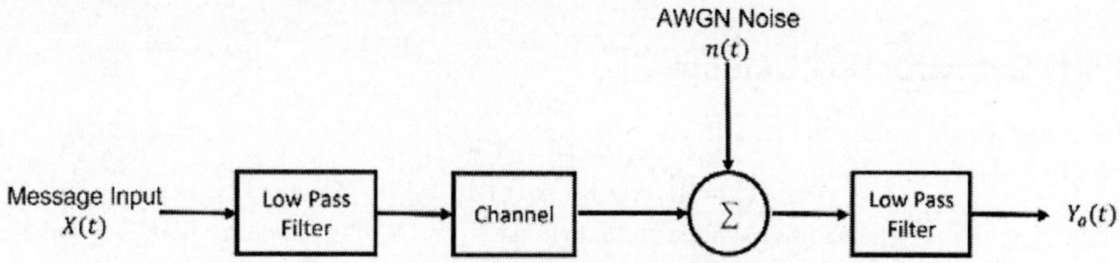

Fig. 2.4 Block diagram of baseband communication system

Fig. (2.4) shows a baseband communication system in which the input signal is transmitted directly without modulating it and is useful for comparing the performance of other systems. In such systems, the receiver is a low pass filter (LPF) of bandwidth W which passes the message signal and rejects each and every frequency component of the noise that does not lie inside the message band.

Considering that $X(t)$ is bandlimited to W with zero mean and power spectral density $S_{XX}(w)$. Given that the channel only introduces additive noise as distortion, the signal at receiver output with time delay t_d is given as

$$X_o(t) = X(t - t_d) \tag{2.5}$$

Thereby, the average output signal power at receiver output is given as

$$\begin{aligned} S_o = E[X_o{}^2(t)] &= E[X^2(t - t_d)] \\ &= \frac{1}{2\pi}\int_{-W}^{W} S_{XX}(w)\,dw = S_X \\ &= S_i \end{aligned} \tag{2.6}$$

where S_i is signal power at receiver input and S_X is average power of signal. Next, the AWGN noise power is given as

$$N_o = E[n_o{}^2(t)] = \frac{1}{2\pi}\int_{-W}^{W} \frac{\eta}{2}\,dw = \frac{\eta W}{2\pi} = \eta B \tag{2.7}$$

where W= $2\pi B$ and B is bandwidth in Hertz. Therefore, the output SNR for baseband communication system is given as

$$\left(\frac{S}{N}\right)_o = \frac{S_i}{\eta B} \tag{2.8}$$

2.5 NOISE IN AMPLITUDE MODULATION SYSTEMS

In amplitude modulation systems, the modulated signal can be demodulated by using a synchronous detector or envelope detector. The modulated signal can be expressed as

$$X_{AM}(t) = A_C[1 + \mu X(t)]\cos\omega_c(t) \tag{2.9}$$

where A_c represents the amplitude of the carrier signal, μ represents the modulation index, and w_c represents the carrier frequency.

2.5.1 Synchronous Detector:

The output signal at the receiver will be

$$Y_0(t) = A_c\mu X(t) + n_c(t)$$

where n_c represents the additive white Gaussian noise (AWGN). Now, let the signal *X(t)* has zero mean then the input signal power can be expressed as

$$S_i = \frac{1}{2}{A_c}^2(1 + \mu^2 S_x) \tag{2.10}$$

Then the signal-to-noise ratio (SNR) at the output of the receiver can be

$$\left(\frac{S}{N}\right)_0 = \frac{\mu^2 S_X}{1+\mu^2 S_X}\left(\frac{S_i}{\eta B}\right) = \frac{\mu^2 S_X}{1+\mu^2 S_X}\gamma \tag{2.11}$$

since $\mu^2 S_X \leq 1$we have

$$\left(\frac{S}{N}\right)_0 \leq \frac{\gamma}{2} \tag{2.12}$$

Hence, (2.12) indicates that the output SNR of the amplitude modulated system is at least 3 dB worse than the DSB and SSB systems.

2.5.2 Envelope Detection:

The amplitude modulated signal is commonly demodulated by using envelope detection. The input to the detector can be expressed as

$$\begin{aligned} Y_i(t) &= X_c(t) + n_i(t) \\ &= \{A_c[1 + \mu X(t)] + n_c(t)\}\cos w_c t - n_s(t)\sin w_c t \end{aligned} \tag{2.13}$$

where $n_s(t)$ denotes quadrature noise component.

1) In case of large SNR: -

when the received SNR is large i.e., $(S/N)_i \gg 1$, $A_c[1 + \mu X(t)] \gg n_i(t)$, and therefore $A_c[1 + \mu X(t)] \gg n_c(t)$ and $n_s(t)$ for all t. Hence, the envelope *V(t)* of the envelope detector can be approximated as

$$V(t) \approx A_c[1 + \mu X(t)] + n_c(t)$$

Therefore, the output SNR can be expressed as

$$\left(\frac{S}{N}\right)_0 = \frac{\mu^2 S_X}{1+\mu^2 S_X}\gamma \tag{2.14}$$

Hence, in the case of large SNR, the performance of the envelope detector is identical to the synchronous detector.

2) In case of small SNR: -

when the received SNR is small i.e., $(S/N)_i \ll 1$ then the envelope of the envelope detector will be dominated by the noise signal. The envelope of the resultant signal can be approximated as

$$V(t) \approx V_n(t) + A_c[1 + \mu X(t)] \cos \phi_n (t) \tag{2.15}$$

where $V_n(t)$ and $\phi_n(t)$ represent the envelope and phase of the noise $n_i(t)$. From Eq. (2.15) one can observe that output does not contain the term directly proportional to *X(t)* and that noise is also multiplicative. Here, the signal *X(t)* is multiplied by noise in the terms of $cos\, \phi_n (t)$, that is random. Therefore, the message signal *X(t)* is multiplied, and the information is lost.

2.6 NOISE IN ANGLE MODULATION SYSTEMS

In angle modulation systems, the modulated signal can be expressed as

$$X_c(t) = \cos[\omega_c t + \phi(t)], \tag{2.16}$$

where,

$$\phi(t) = \begin{cases} K_p X(t) \; for \; PM \\ K_f \int_{-\infty}^{t} X(t) \; for \; FM \end{cases} \tag{2.17}$$

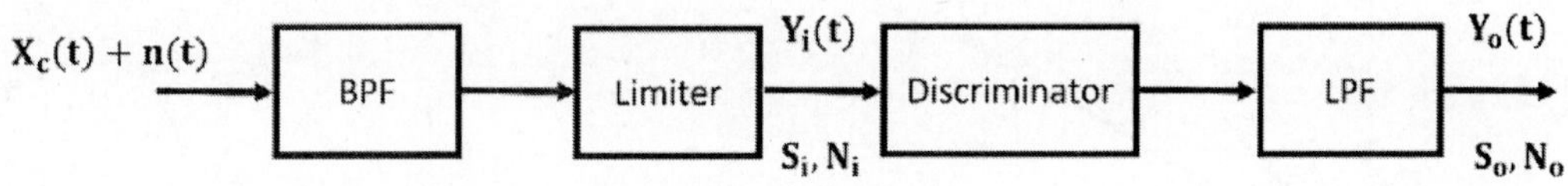

Fig. 2.5 Angle modulation system

Block diagram of an angle modulation system is shown in Fig. (2.5). The bandwidth B_T of the prediction filter is given by 2(D+1) B, where B is the bandwidth of the message signal and D is the deviation ratio. The detector input is

$$\begin{aligned} Y_i(t) &= X_c(t) + n_i(t) \\ &= \cos\big(\omega_c(t) + \phi(t)\big) + n_i(t). \end{aligned} \tag{2.18}$$

The carrier amplitude remains constant, therefore

$$S_i = E[X_i^2(t)] = \frac{1}{2} A_c^2, \tag{2.19}$$

and,

$$N_i = \eta B_T, \tag{2.20}$$

hence,

$$\frac{S_i}{N_i} = \frac{A_c^2}{2\eta B_T}, \tag{2.21}$$

which is independent of X(t).

The $\left(\frac{S}{N}\right)_i$ of Eq. (2.21) is called signal-to-noise-ratio (SNR).

Because $n_i(t)$ is the narrowband, we write

$$n_i(t) = v_n(t) \cos[\omega_c t + \phi(t)]. \tag{2.22}$$

where $v_n(t)$ is Rayleigh distributed and $\phi_n(t)$ is uniformly distributed in $(0, 2\pi)$. Then $Y_i(t)$can be written as

$$Y_i(t) = V(t) \cos[\omega_c t + \theta(t)], \tag{2.23}$$

where $V(t) = \{[A_c \cos\phi + v_n(t)\cos\phi_n(t)]^2 + [A_c \sin\phi + v_n(t)\sin\phi_n(t)]^2\}^{\frac{1}{2}}$ and

$$\theta(t) = \tan^{-1}\left(\frac{A_c \sin\phi + v_n(t)\sin\phi_n(t)}{A_c \cos\phi + v_n(t)\cos\phi_n(t)}\right)$$

The amplitude variation V(t) is suppressed by the limiter and SNR is derived from $\theta(t)$. The expression for $\theta(t)$ is complicated for analysis. Therefore, let us consider the detector to be ideal and its output is given by

$$Y_o(t) = \begin{cases} \theta(t)\ for\ PM, \\ \dfrac{d\theta(t)}{dt}\ for\ FM. \end{cases} \tag{2.24}$$

2.6.1 Signal Dominance Case:

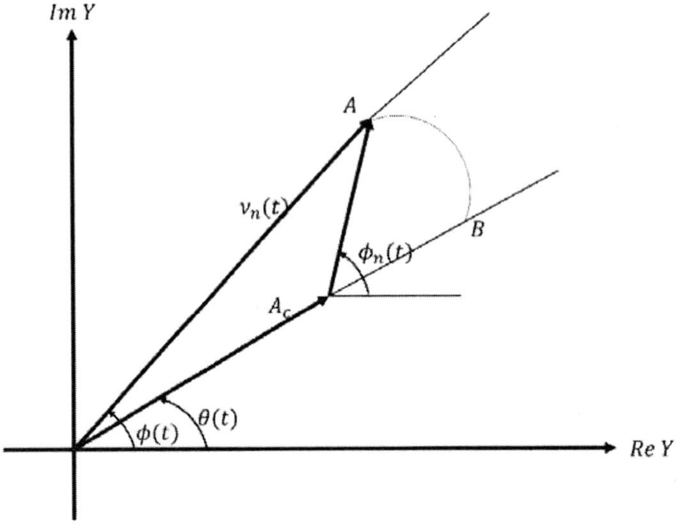

Fig. *2.6* Phasor diagram

$Y_i(t)$ can be written in the passband form as

$$Y_i(t) = Re\left[Y(t)e^{j\omega_c(t)}\right], \tag{2.25}$$

where

$$Y(t) = A_c e^{j\phi(t)} + v_n(t)e^{j\phi_n(t)}. \tag{2.26}$$

and $V_n(t) \ll A_c$ for all t. The length of arc AB is given from Fig. (2.6) as

$$L = Y(t)[\theta(t) - \phi(t)], \tag{2.27}$$

hence

$$\theta(t) \approx \phi(t) + \frac{v_n(t)}{A_c}\sin\phi_n(t) \tag{2.28}$$

$$= \phi(t) + \frac{n_s(t)}{A_c}.$$

From Eq. (2.24) and (2.17) the detector output is

$$Y_o(t) = \theta(t) = K_p X(t) + \frac{n_s(t)}{A_c}\ \text{for PM.} \tag{2.29}$$

$$Y_o(t) = \frac{d\theta(t)}{dt} = K_f X(t) + \frac{n_s'(t)}{A_c} \text{ for FM.} \tag{2.30}$$

1) Output SNR $\left(\frac{S}{N}\right)_o$ in PM systems:

From Eq. (2.29)

$$S_o = E[K_p^2 X^2(t)] = K_p^2 E[X^2(t)] = K_p^2 S_X. \tag{2.31}$$

$$N_o = E\left[\frac{1}{A_c^2} n_s^2(t)\right] = \frac{1}{A_c^2} E[n_s^2(t)] = \frac{1}{A_c^2}(2\eta B). \tag{2.32}$$

Hence,

$$\left(\frac{S}{N}\right)_o = \frac{K_p^2 A_c^2 S_X}{2\eta B}. \tag{2.33}$$

From Eq. (2.19)

$$\gamma = \frac{S_i}{\eta B} = \frac{A_c^2}{2\eta B}. \tag{2.34}$$

Then Eq. (2.34) becomes

$$\left(\frac{S}{N}\right)_o = K_p^2 S_X \gamma\,. \tag{2.35}$$

2) Output SNR $\left(\frac{S}{N}\right)_o$ in FM systems:

From Eq. (2.31)

$$S_o = E[K_f^2 X^2(t)] = K_f^2 E[X^2(t)] = K_f^2 S_X \tag{2.36}$$

$$N_o = E\left[\frac{1}{A_c^2}[n_s'(t)]^2\right] = \frac{1}{A_c^2} E[[n_s'(t)]^2] = \frac{2}{3}\frac{\eta}{A_c^2}\frac{W^3}{2\pi}, \tag{2.37}$$

hence,

$$\left(\frac{S}{N}\right)_o = \frac{3\pi A_c^2 K_f^2 S_X}{\eta W^3}. \tag{2.38}$$

Using Eq. (2.34)

$$\left(\frac{S}{N}\right)_o = 3\left(\frac{K_f^2 S_X}{W^2}\right)\left(\frac{A_c^2}{2\eta B}\right) = 3\left(\frac{K_f^2 S_X}{W^2}\right)\gamma \tag{2.39}$$

Since $\Delta\omega = |K_f X(t)|_{max} = K_f[|X(t)| \ll 1]$, Eqn. (24) is given by

$$\left(\frac{S}{N}\right)_o = 3\left(\frac{\Delta\omega}{W}\right)^2 S_X\gamma = 3D^2 S_X \gamma, \tag{2.40}$$

where D is the deviation ratio.

2.7 PRE-EMPHASIS

As we have seen earlier that the ability of FM to suppress noise decreases as the frequency is increased. Thus, by increasing the amplitude of only the high frequency components of the message signal to be transmitted before modulation at the transmitter side is termed as pre-emphasis. Pre-emphasis circuit as shown in Fig. (2.7), increases the energy content of the selected higher-frequency components such that they are stronger compared to the noise components which also have high frequency. This process improves the signal to noise ratio and increases overall intelligibility and fidelity of the communication system.

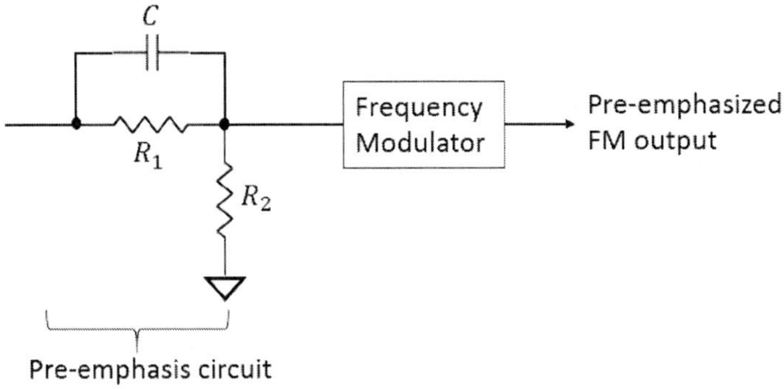

Fig. 2.7 Pre-emphasis circuit

As can be seen from Fig. (2.7) pre-emphasis circuit is a high pass filter or we can say that it acts as a differentiator which allows higher-frequencies to pass through and block the lower frequencies. The 3dB cut off frequency for pre-emphasis circuit can be computed as:

$$f_l = \frac{1}{2\pi RC}$$

The pre-emphasis circuit has upper break frequency denoted as f_u in Fig. (2.8). After f_u the signal enhancement flattens out as is evident from the characteristic curve. The upper break frequency can be computed as:

$$f_u = R_1 + \frac{R_2}{2\pi R_1 R_2 C}$$

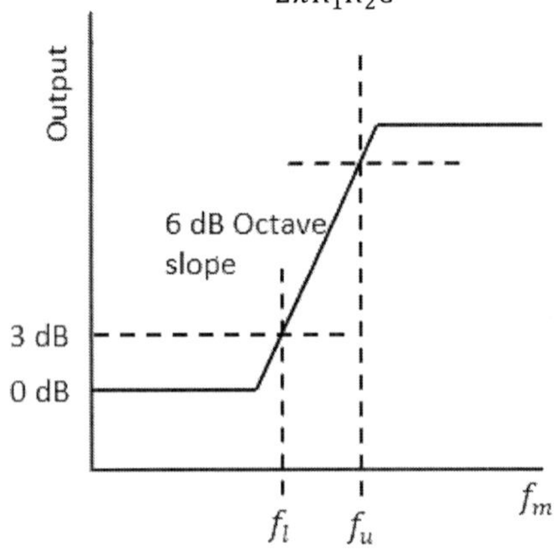

Fig. 2.8 Pre-emphasis circuit characteristics

2.8 DE-EMPHASIS

In the de-emphasis circuit, the amplitude level of the high frequency signal received by the receiver is reduced by the same amount as was increased in pre-emphasis at the transmitter side to bring them to their original amplitude levels. Analogous to the pre-emphasis process, de-emphasis circuit is employed at the receiver side after signal is received. As can be seen from Fig. (2.9) de-emphasis circuit is a low pass filter or we can say that it acts as an integrator which allows lower frequencies to pass through and block the higher frequencies.

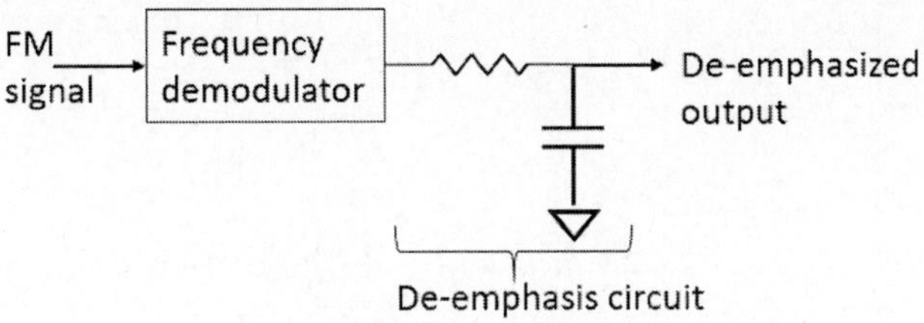

Fig. 2.9 De-emphasis circuit

2.9 THRESHOLD EFFECT IN ANGLE MODULATION

The phasor is given as

$$Y_i(t) = \text{Re}[Y(t)e^{j\omega_c t}] \tag{2.41}$$

where

$$Y(t) = A_c e^{j\phi(t)} + v_n(t)e^{j\phi_n(t)} \tag{2.42}$$

In Eq. (2.42), when $A_c^2 \ll E[n_i^2(t)]$, the term $v_n(t)e^{j\Phi_n(t)}$ will dominant. For this case, the phase of detector input is given as

$$\theta(t) \approx \phi_n(t) + \frac{A_c}{v_n(t)} Sin[\Phi(t) - \phi_n(t)], \tag{2.43}$$

In the above case the message signal has been corrupted by the noise. As a result, the possibility of recovering of message signal is very less because of higher noise. The threshold effect is represented in Fig (2.10).

When $A_c^2 \gg E[n_i^2(t)]$, $v_n(t) \ll A_c$, which results in phasor $Y(t)$ to trace randomly around the end of the carrier phase. However, as the noise increases, the tip of the resulting phasor $Y(t)$ may drift away from the carrier phase's terminus and may occasionally encircle the origin, where the phase changes by 2π very rapidly [Fig. (2.11)]. Whenever an encirclement occurs, a spike will appear in the frequency discriminator's output.

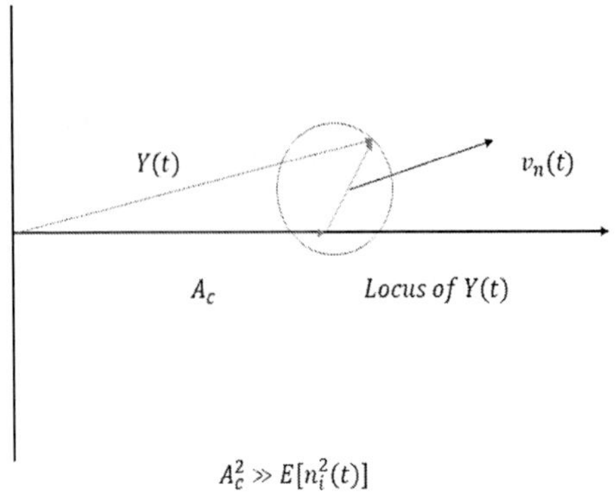

Fig. 2.10 Threshold effect in FM system for $A_c^2 \gg E[n_i^2(t)$

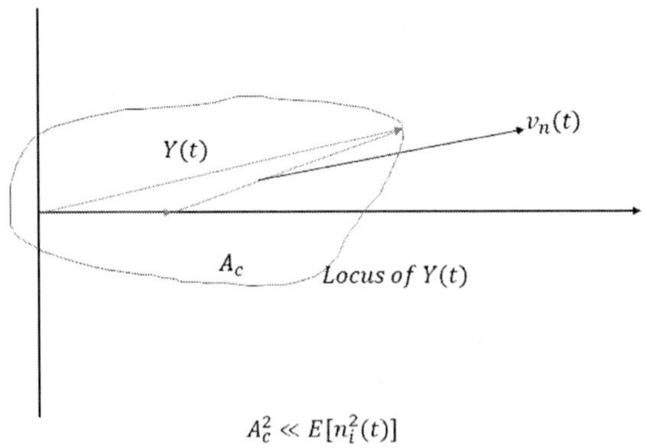

Fig. 2.11 Threshold effect in FM system for $A_c^2 \ll E[n_i^2(t)$

Fig. 2.12

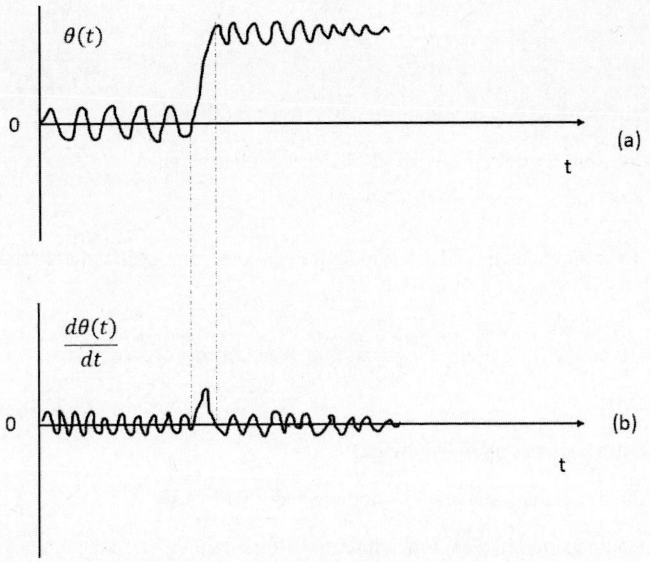

Threshold effect in angle modulation

PRACTICAL

Experiment: -1 Measurement of noise power and noise power spectral density.

Objectives: to measure the power spectral density of PN sequence noise signal

Theory:

Noise power: just give the noise signal as input to multiplier and choose appropriate integrating time.

Signal power: same as above

to calculate PSD: major noise power by keeping the integrating time at Max of 15 seconds.

bandwidth of sources 2 megahertz

single sided PSD is equal to noise power upon bandwidth

and not is equal to usually double-sided PSD equal to single sided PSD by two

signal to noise ratio is calculated using different procedure for different modulation techniques

Equipment's:

- ACL 05
- power supply
- 20 MHz dual trace was low scope

Note: Keep all switches fault in off position

Procedure:

1) Connect the power supply with power polarity to kit ACL 05 and switch it on.

2) connect the CLK out and CLK in post off my generator connect outpost of noise generator to noise in post of header.

3) Keep the label part P1 of noise generator to maximum position.

4) Keep the attenuation part P42 minimum position.

5) Connect outpost officer in post of power meter.

6) Keep timer pot P5 to maximum position.

7) Major noise power PNI on power meter.

9) Single sided PSD noise power up on bandwidth and not equal to single sided PSD upon.

Calculations:

Single sided PSD = Noise power/ Bandwidth

Noise power spectral density = Single sided PSD/2

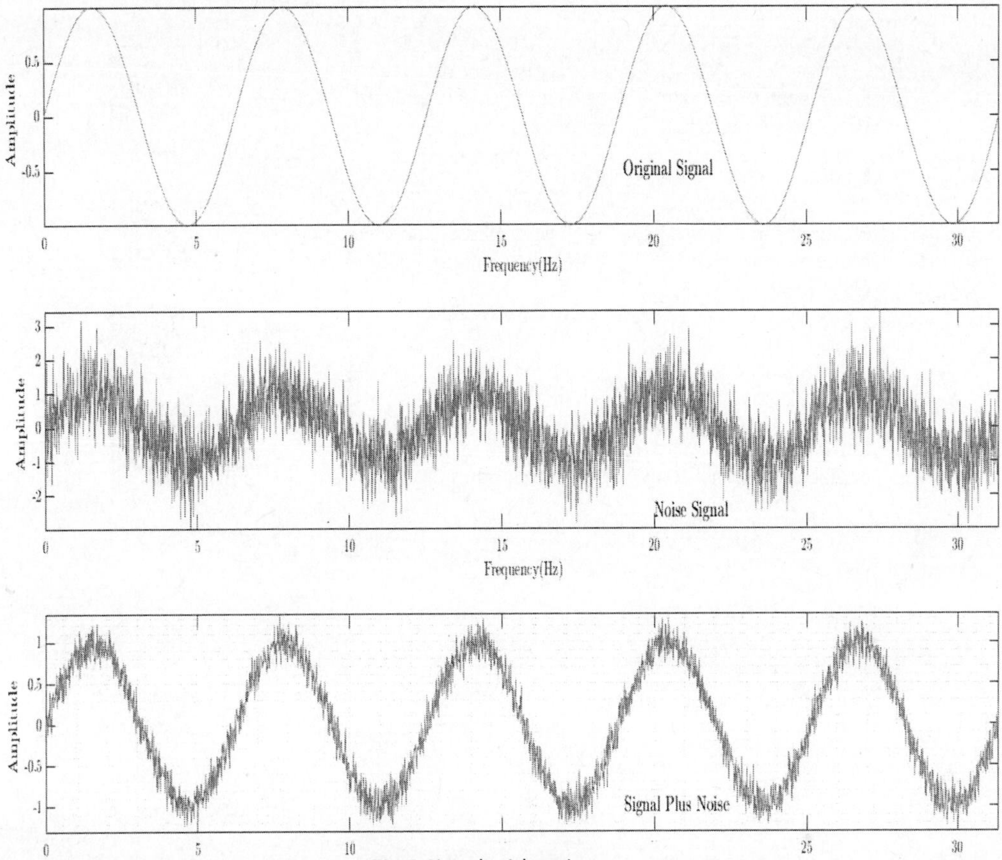

Fig.1 Signal with noise

References and Suggested Reading

8. B. P. Lathi, and Zhi Ding, "Modern Digital and Analog Communication Systems", Oxford University Press, 2010.

9. Simon Haykin, "Communication Systems", Wiley, 2007.

10. Herbert Taub, Donald L. Schilling, and Gautam Saha, "Principal of Communication Systems", McGraw Hill Education, 2017.

11. P.F. Panter, "Modulation, Noise and Spectral Analysis", McGraw-Hill, New York, 1965.

12. N Abramson, "Bandwidth and Spectra Phase- and Frequency-Modulated Waves",IEEE Transaction Communication System, CS_11, PP. 407-414, 1963.

13. https://nptel.ac.in/courses/117102059

14. https://ocw.mit.edu/courses/16-36-communication-systems-engineering-spring-2009/pages/lecture-notes/

3 PULSE MODULATION

UNIT SPECIFICS

Through this unit we have discussed the following aspects:

- *Review of pulse modulation and sampling process.*
- *Study of pulse amplitude and pulse code modulation,*
- *Differential Pulse code modulation,*
- *Delta modulation.*
- *Noise consideration in PCM,*
- *Digital Mutiplexers.*

The practical applications of the topics are discussed for generating further curiosity and creativity as well as improving problem solving capacity.

Besides giving a large number of multiple choice questions as well as questions of short and long answer types marked in two categories following lower and higher order of Bloom's taxonomy, assignments through a number of numerical problems, a list of references and suggested readings are given in the unit so that one can go through them for practice. It is important to note that for getting more information on various topics of interest some QR codes have been provided in different sections which can be scanned for relevant supportive knowledge.

After the related practical, based on the content, there is a "Know More" section. This section has been carefully designed so that the supplementary information provided in this part becomes beneficial for the users of the book. This section mainly highlights the initial activity, examples of some interesting facts, analogy, history of the development of the subject focusing the salient observations and finding, timelines starting from the development of the concerned topics up to the recent time, applications of the subject matter for our day-to-day real life or/and industrial applications on variety of aspects, case study related to environmental, sustainability, social and ethical issues whichever applicable, and finally inquisitiveness and curiosity topics of the unit.

RATIONALE

This unit on pulse modulation helps students to get a primary idea about the various pulse modulation techniques. We have first started with the sampling process, where we have introduced sampling rate. Then, we have covered Pulse Amplitude and Pulse code modulation (PCM), Differential pulse code modulation

(DPCM), Delta modulation (DM) in a detailed manner with the help of block diagrams. Then we have covered the noise considerations in PCM. After this, we have explained, digital multiplexers and different multiplexing techniques such as frequency division multiplexing (FDM) and time division multiplexing (TDM).

Pulse modulation is a type of analogue modulation technique which is employed on discrete information signals. The continuous signals which are obtained from nature can be converted to discrete signals with the help of sampling as is explained in this chapter.at the end of this chapter we will understand the advantages of pulse modulation, various effects which deteriorate the performance and then later on as we progress we will learn about the ways to mitigate them.

PRE-REQUISITES

Mathematics: Calculus (Class XII)

Physics: Basic of Communication (Class XII)

UNIT OUTCOMES

List of outcomes of this unit is as follows:

U1-O1: Describe basics of pulse modulation and review sampling process.

U1-O2: Explain pulse amplitude and pulse code modulation

U1-O3: Describe differential pulse code modulation

U1-O4: Explain noise consideration in PCM

U1-O5: Explain digital multiplexers

This table below need to be filled after discussion as marked with red.

Unit-1 Outcomes	EXPECTED MAPPING WITH COURSE OUTCOMES (1- Weak Correlation; 2- Medium correlation; 3- Strong Correlation)					
	CO-1	CO-2	CO-3	CO-4	CO-5	CO-6
U1-O1	3	3	3	-	3	1
U1-O2	1	1	2	2	1	-
U1-O3	2	1	3	1	2	1
U1-O4	-	-	3	1	2	2
U1-O5	3	3	3	-	3	1

3.1 SAMPLING

Sampling is the process of measuring the instantaneous values of a continuous time domain signal in discrete signal. The continuous time signal x(t) is transformed into a sampled signal, $x_s(t)$, as shown in Fig. (3.1)

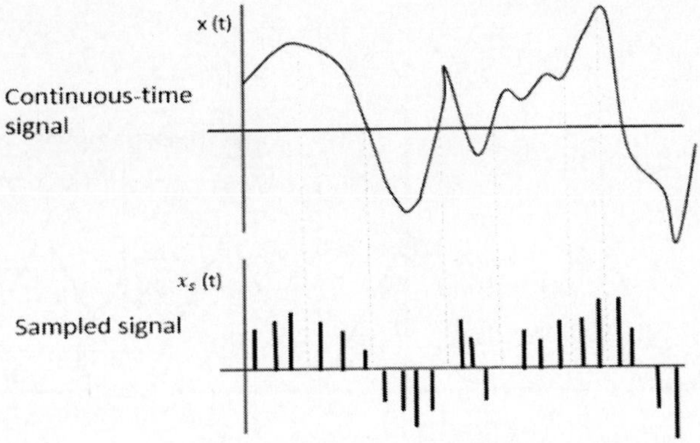

Fig 3.1 Continuous time and sampled signal

While sampling the signals the gap between the samples is always fixed. This gap is called sampling period T_s.

$$\text{Sampling Frequency } (F_s) = \frac{1}{T_s}$$

Where, T_s is sampling duration and F_s is sampling frequency or the sampling rate.

1) When the sampling signal frequency F_s is larger than or equal to the input signal frequency F_m, the discrete form of a signal can be restored to its original form using samples or the sampled (discrete) signal.

$$\mathbf{F_s \geq 2F_m}$$

2) When the sample frequency (F_s) is twice as high as the input signal frequency, the Nyquist criteria for Sampling are satisfied (F_m). The sampling frequency is equal to the input signal frequency divided by the Nyquist rate.

$$\mathbf{F_s = 2F_m}$$

3) Aliasing effect occurs when the sampling frequency (F_s) is less than twice the frequency of the input signal.

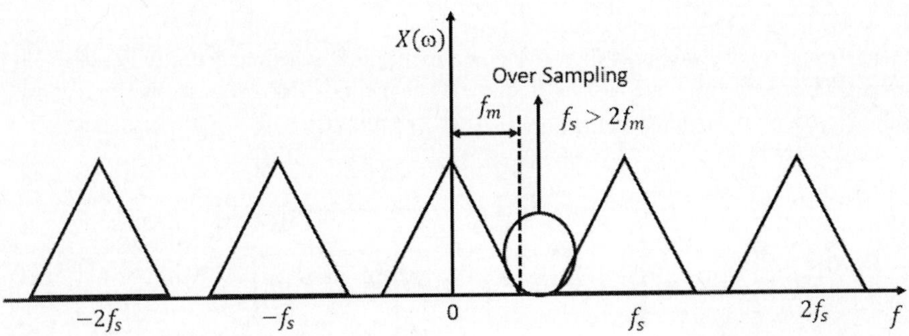

Fig 3.2 (a) Signal reconstruction with over sampling

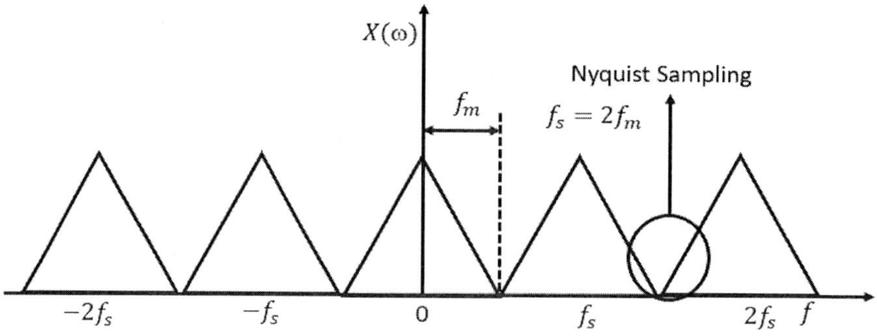

Fig 3.2 (b) Signal reconstruction with Nyquist sampling

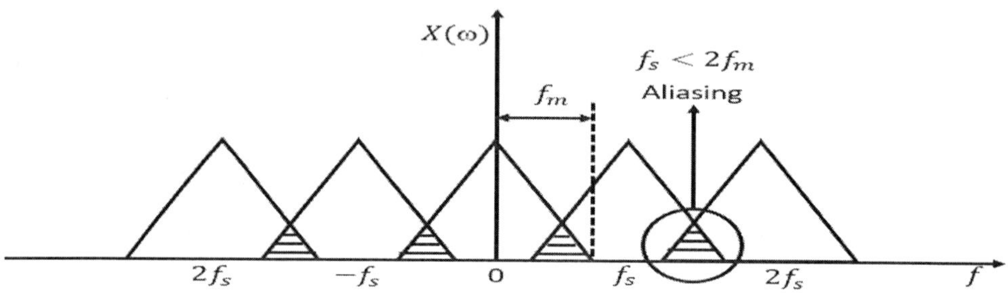

Fig 3.2 (c) Signal reconstruction with aliasing effect

3.1.1 Low Pass Filter

The frequency response of a low-pass filter (LPF) is a rectangular function, and it fully rejects all frequencies above the cutoff frequency w_o while passing frequencies below it. A RC low pass filter to reconstruct signal is shown below in Fig. (3.2). Here the cutoff frequency w_o of LPF is $\frac{1}{RC}$

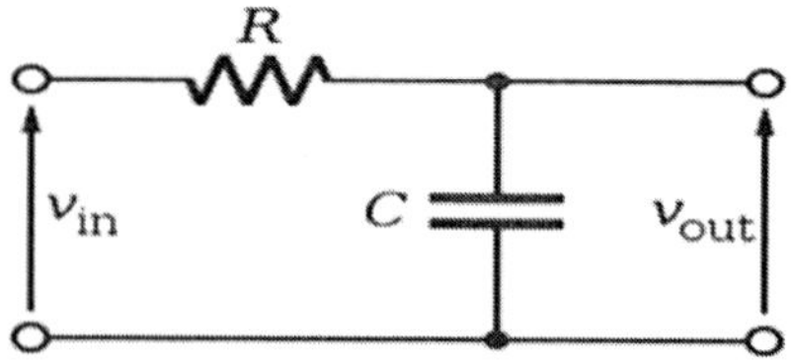

Fig. 3.3 RC low pass filter

The LPF is used for the reconstruction of the original or desired signal.

3.2 PULSE MODULATION

Pulse modulation is a type of modulation in which a train of pulses is used as the carrier wave. One or more of its parameters, such as amplitude, width and position are modified in order to carry the information contained in the message signal.

3.2.1 Pulse Amplitude Modulation

The amplitude of the pulse carrier swings proportionally to the instantaneous amplitude of the message signal in PAM, an analogue modulation method. The pulse amplitude modulated signal will match the original signal's amplitude as it travels along the entire wave's path. By putting a signal through an effective Low Pass Filter (LPF) with a precise cut-off frequency, Natural PAM can reconstitute a signal sampled at the Nyquist rate. The Pulse Amplitude Modulation is depicted in the following figures.

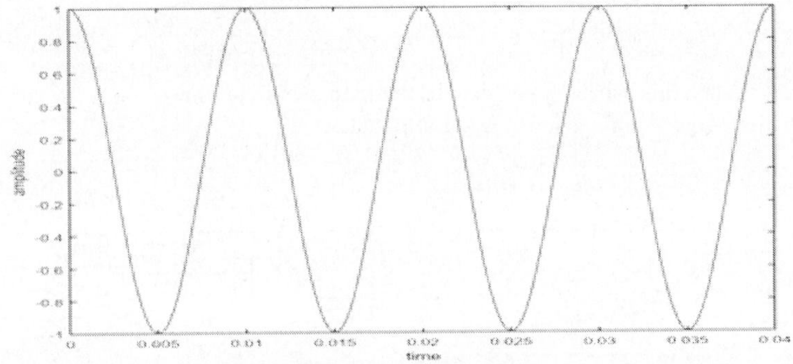

Fig 3.4 Message Signal

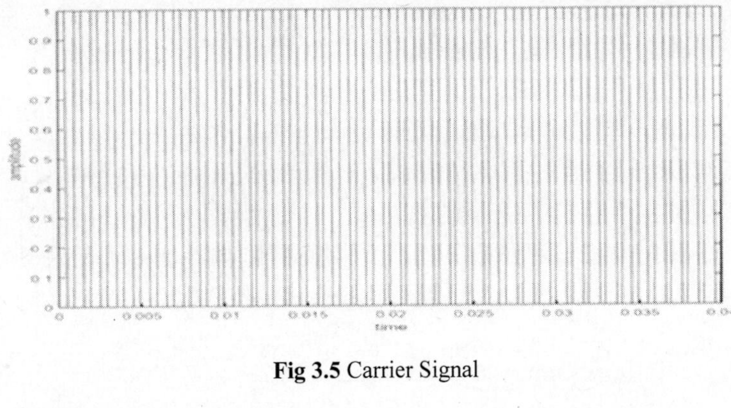

Fig 3.5 Carrier Signal

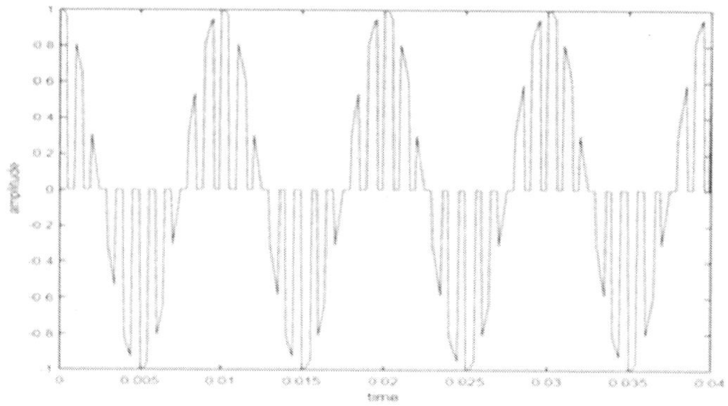

Fig 3.6 Pulse Amplitude Modulation Signal

3.2.2 Pulse Width Modulation

PWM is an analogue modulation scheme by which the duration, width, or time of the pulse carrier is varied proportionally to the message signal's instantaneous amplitude.

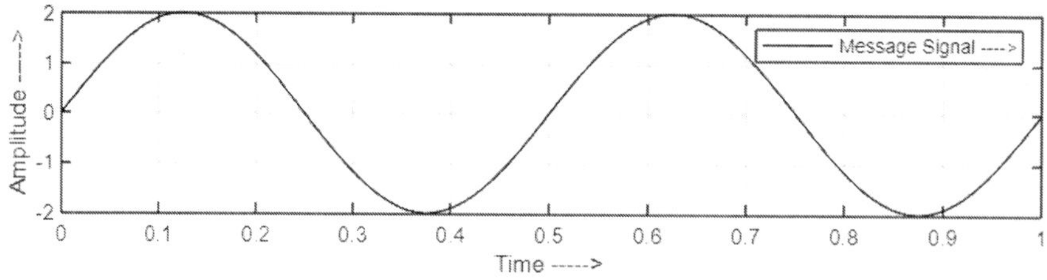

Fig 3.7 Message signal

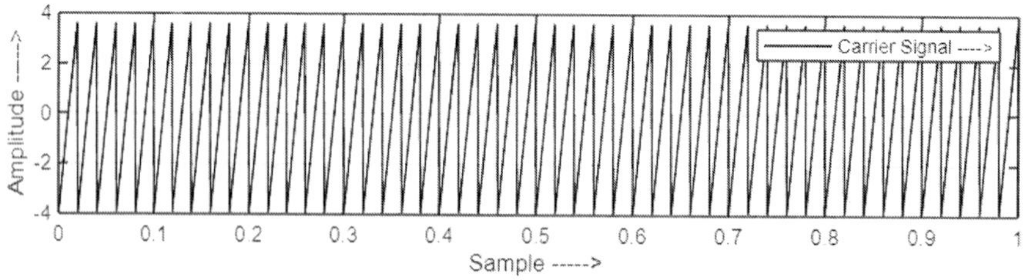

Fig 3.8 Carrier Signal

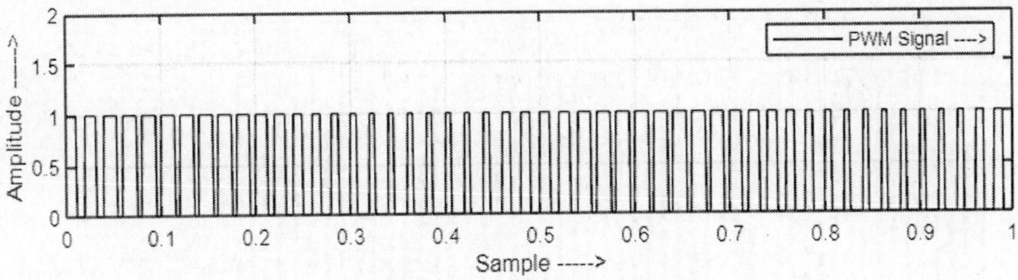

Fig 3.9 Pulse Width Modulated Signal

3.2.3 Pulse Position Modulation

Pulse Position Modulation is a modulation technique that allows for variations in the position of the pulses based on the amplitude of the sampled modulating signal (PPM). It is an another kind of modulation technique in which the amplitude and duration of the pulses remain constant while only the place of the pulses varies. The position of each pulse varies according to the message signal's instantaneous sampled value.

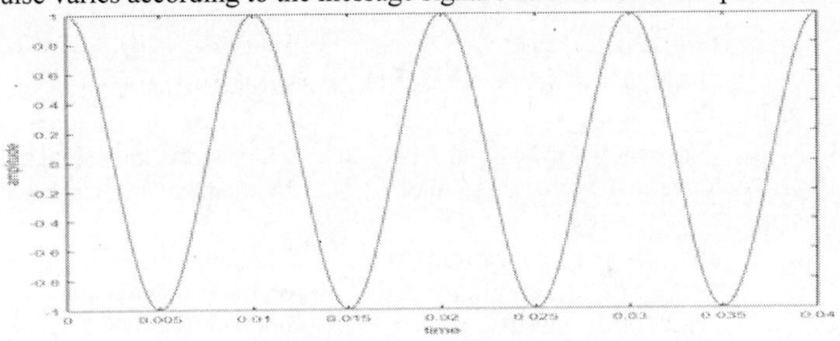

Fig 3.10 Message Signal

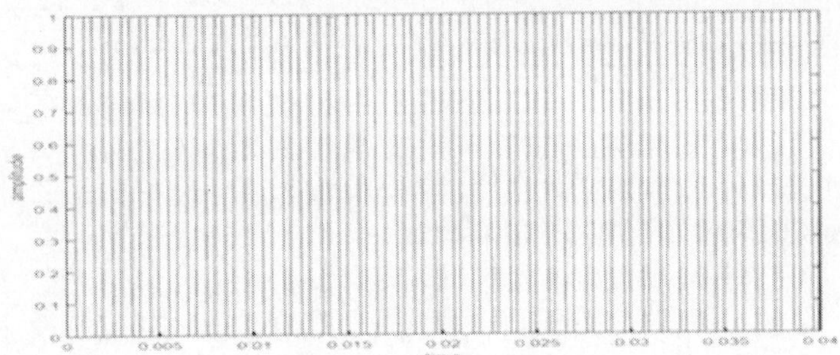

Fig 3.11 Carrier Signal

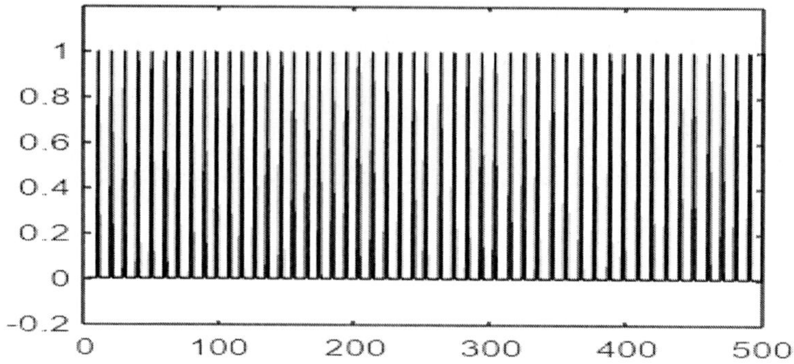

Fig 3.12 Pulse Position Modulated Signal

Difference between different modulation schemes are summarized as:

PAM	PWM	PPM
Amplitude is varied	Width is varied	Position is varied
The bandwidth of a pulse is governed by its width.	The bandwidth of a pulse is governed by its rising time.	The bandwidth is governed by the pulse's rising time.
Instantaneous power transmitted varies with the amplitude of the pulses.	Instantaneous power transmitted varies with the amplitude and width of the pulses	The pulse width has no effect on the instantaneous transmitter power.
High system complexity	System complexity is low	System complexity is low
Noise interference is high	Noise interference is low	Noise interference is low
It is similar to amplitude modulation	It is similar to frequency modulation	It is similar to phase modulation

3.2.4 Demodulation of PAM/PWM/PPM signals

PAM signals can be demodulated by passing through a LPF which retains the low frequency message signal and smoothens out the pulse train information.

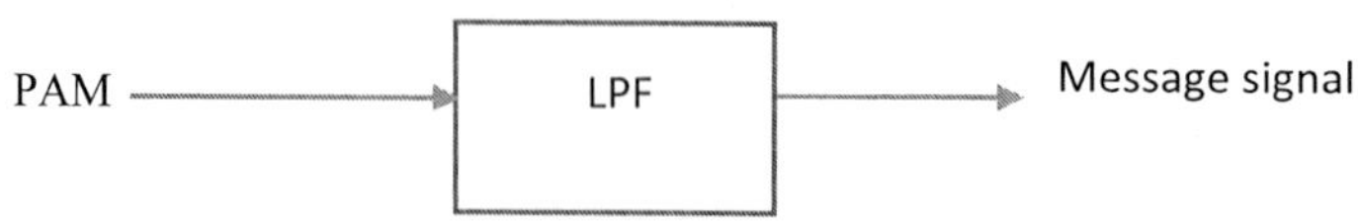

PWM/PPM are demodulated by first converting them into PAM signal and then perform low pass filtering.

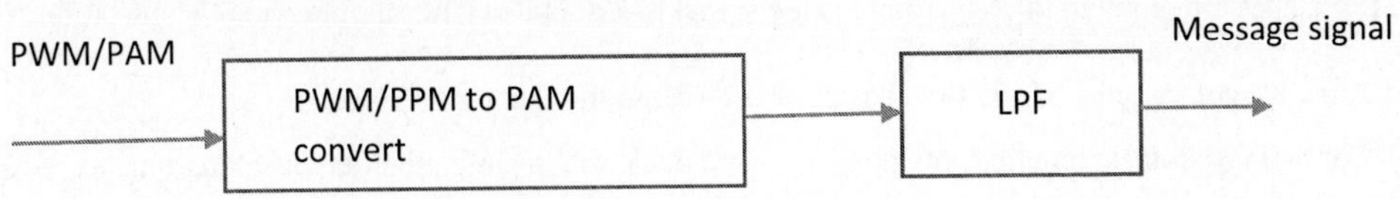

After having discussion of Pulse analog modulation, we will discuss about digital pulse modulation. It is of four type-

1. Pulse Code Modulation (PCM)
2. Differential Pulse Code Modulation (DPCM)
3. Delta Modulation (DM)
4. Adaptive Delta Modulation (ADM)

3.2.5 Pulse code Modulation

A message signal m(t) is represented by a series of encoded pulses, which is achieved by discretizing the signal in time and amplitude in PCM. This technology is revolutionary for the transition from analog to digital communication. The time parameter in PAM is discrete while the amplitude stays continuous. In this manner, PAM and PCM are distinct.

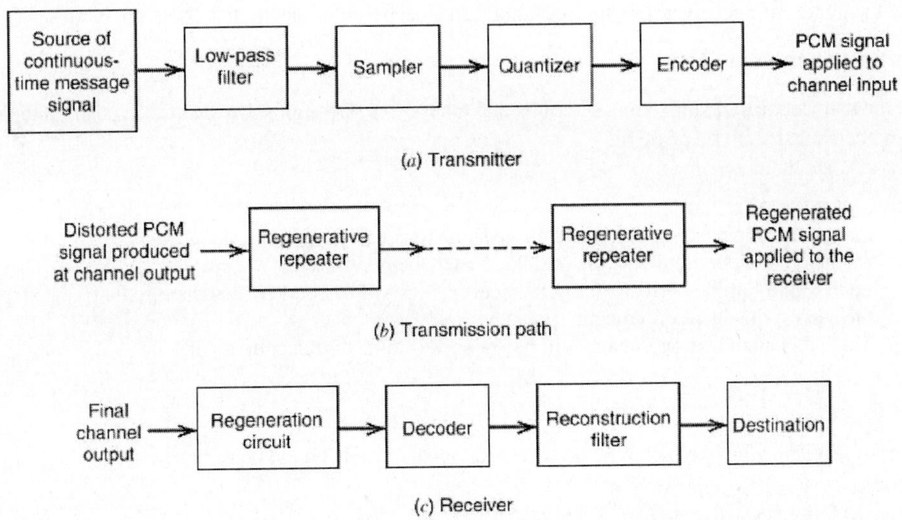

Fig. 3.13 PCM Block Diagram

3.2.6 Working of PCM system

Through sampling and quantization, an analog signal may be transformed into a digital signal. First the message signal is sampled, and then the value of the amplitude of each sample is rounded off to the closest as quantization levels. The quantisation levels are always finite in number. Now the signal is discrete in both time and amplitude. Hence, two stages are required to convert an analog signal to a digital signal: sampling and quantization

- First, we get samples of this signal as par by the sampling theorem.
- For taking samples, time instants $t_0, t_1, t_2 \cdots$ are marked at equal time intervals along time axis.
- The magnitude of the signal is measured at each of these time instants, and these are referred to as samples.
- Signal is now defined only at samples and hence, the signal is discrete. But since the magnitude still can take any value, hence the signal is still analog. This problem is solved by quantization. In quantization, we divide the total amplitude range of the signal into a finite number of permissible amplitude levels.

For e.g., $x(t) = V_m \sin(\omega t)$
The amplitude of $x(t)$ varies from $-V_m$ to V_m

If it is partitioned into L intervals, each having magnitude $\Delta = \frac{V_{\max}-V_{\min}}{L} = \frac{2V_m}{L}$, then it is said to be quantized. Here Δ is called as step size.
The transmitter consists of mainly three operations-sampling, quantizing and encoding. Sampling is done at Nyquist rate or higher which results in discrete time signal. A single circuit can be utilised to perform both the quantisation and encoding operations. This circuit is known as an analog to digital convertor (ADC).

Regeneration of degraded signals, decoding, and demodulation of a train of quantized samples are the three fundamental activities of a receiver. Digital to analog (DAC) converters accomplish these operations.

Mathematical Analysis of PCM

Having all the knowledge of quantization, sampling and other PCM operations, we are now in position to discuss about the mathematical analysis of PCM.

1. First, the complete dynamic range of the signal is partitioned into L equal steps.
2. As the quantization voltage, the middle of each step is chosen.
3. Each of the sample corresponding to specific step will be round off to the middle of the step or to the nearest quantization voltage.
4. Then, this quantization voltage will be encoded into its digital equivalent.

Note

The number of levels in which dynamic range will be divided depends on the no of bits, n used for encoding.

$$L = 2^n$$

$$n = 2\,; \quad L = 4 \quad \text{And so on.}$$

Example 1: Let input have dynamic range from V_{min} to V_{max} and let n number of bits are used for encoding.

Then no of levels, $L = 2^n$

Then step size, $\Delta = \frac{V_{max}-V_{min}}{2^n}$

Quantization error $(\epsilon_{max}) = \pm\frac{\Delta}{2}$

For instance, take $m(t) = 2sin\omega t \; and \; n = 3$

Then L= 8 levels

Step size, $\Delta = \frac{2-(-2)}{8} = 0.5$

Quantization error $(\epsilon_{max}) = 0.25$

Bandwidth of PCM system

Now, we will evaluate the transmission bandwidth for PCM system. Let us assume that quantizer uses n number of binary digits to represent each level.

Let encoded PCM signal uses n-bits/ sample for encoding.

If T_b = Bit duration of 1 bit

$T_s = Sample\ Time$

Then, $T_s = nT_b$

$r_b = nf_s$

Bandwidth of PCM system= $\frac{r_b}{2} = \frac{nf_s}{2}$

3.3 DIFFERENTIAL PULSE CODE MODULATION (DPCM):

A technique of transforming a digital signal from an analogue signal is called DPCM. In this method, the analog signal is sampled, the distinction between the sampled value and the anticipated value is quantized, and finally the signal is encoded to create a digital value.

3.3.1 DPCM Transmitter

The sampled signal's abbreviation is $x(nT_S)$ and the anticipated signal is shown by $\hat{x}(nT_S)$. The comparator calculates the disparity between the true value x(nTs) and the anticipated value $\hat{x}(nT_S)$. This is called error in signal and is denoted as $e(nT_S)$ [2].

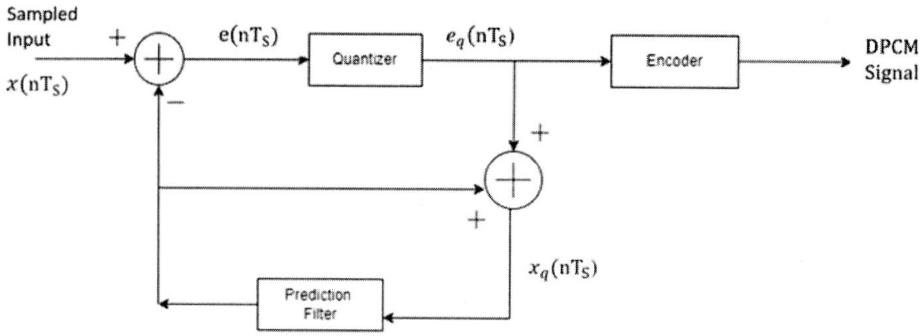

Fig 3.14 DPCM Transmitter

$$e(nT_S) = x(nT_S) - \hat{x}(nT_S) \quad (3.2)$$

The output of quantizer is given by $e_q(nT_S)$, where previous anticipated value is added and given as input to the prediction filter, denoted by $x_q(nT_S)$.
This brings the anticipated and the sampled signal closer together. The quantized error signal $e_q(nT_S)$ is very small and can be encoded by using a small number of bits. Thus, the number of bits per sample is reduced in DPCM [2].

The output of the quantizer would appear as,

$$\boldsymbol{e_q(nT_S)} = e(\mathbf{nT_S}) + q(\mathbf{nT_S}) \quad (3.2)$$

Here $q(nT_S)$ is quantization error. From fig (3.14), the prediction filter input $x_q(nT_S)$ is the sum of $\hat{x}(nT_S)$ and the quantizer output $e_q(nT_S)$.

$$x_q(nT_S) = \hat{x}(nT_S) + e_q(nT_S) \quad (3.3)$$

By using as a substitute for e_q(nTs) from the equation (3.2) in equation (3.3) we get,

$$x_q(nT_S) = \hat{x}(nT_S) + e(nT_S) + q(nT_S) \quad (3.4)$$

Equation (3.1) can written as,

$$e(nT_S) + \hat{x}(nT_S) = x(nT_S) \quad (3.5)$$

from the above equations (3.4) and (3.5) we get,

$$\boldsymbol{x_q(nT_S)} = x(\mathbf{nT_S}) + \mathbf{q(nT_S)} \quad (3.6)$$

Therefore, the sum of original sample value and quantized error q(nT$_S$) gives the quantized signal x_q(nT$_S$). Both positive and negative quantized errors are possible. Therefore, the prediction filter's output is independent of its properties [2].

3.3.2 DPCM Receiver

In order to reassemble the digital signal that was received, the DPCM receiver consists of a decoder and prediction filter.

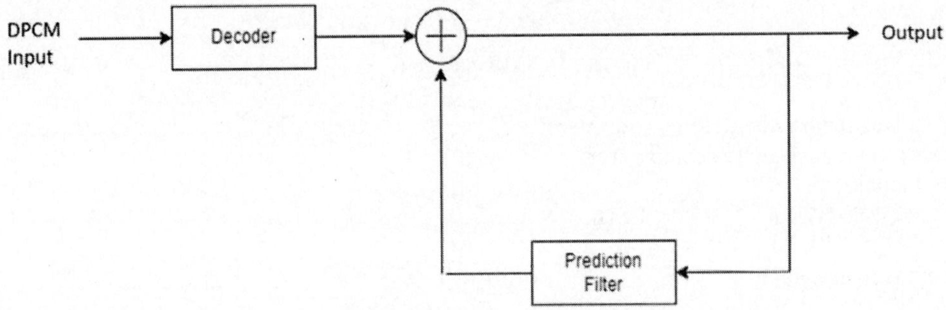

Fig 3.15 DPCM Receiver

Using the previous outputs, the predictor undertakes a value. To produce a better output, the decoder's input is processed, and its output is combined with the predictor's output. This indicates that the decoder will first recreate the original signal's quantized form. As a result, the difference between the signal at the receiver and the true signal is quantization error q(nT$_S$), which is permanently incorporated into the signal after reconstruction.

3.4 DELTA MODULATION

When signal quality is not the main concern, another analogue to digital conversion technique called a delta modulation is performed. It is the most basic type of differential pulse code modulation, where samples is encoded into a data stream of n bits, but delta modulation reduces it to a data stream of just one bit. The signal correlation is increased as the sampling interval is decreased. In Delta Modulation, sampled error signal is transmitted instead of sampled message signal. This reduces the information bits hence only a change in the signal's amplitude from the previous sample i.e. error signal is transmitted. If there is no change, the modulated signal remains the same, either 0 or 1. The analogue signal must be sampled at a rate several times greater than the Nyquist rate in order to delta modulation to obtain high signal to noise ratios.

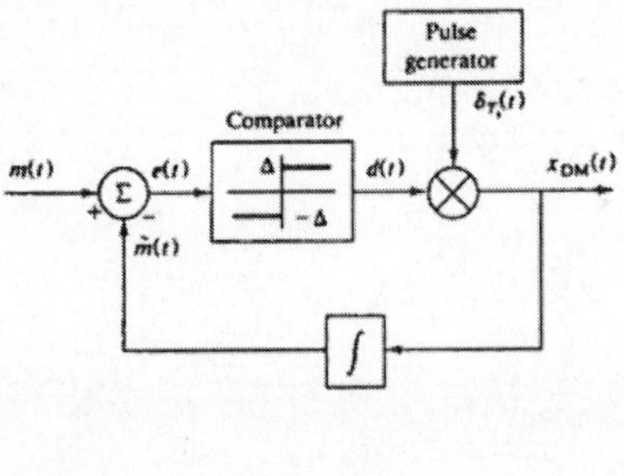

Fig 3.16 Block diagram of delta modulation

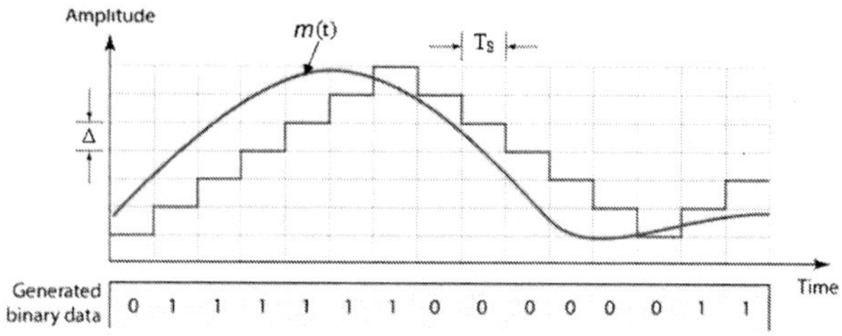

Fig 3.17 Delta Modulation

The input to the comparator is, $e(t) = m(t) - \tilde{m}(t)$
where, $e(t)$ = input signal to comparator
$m(t)$ = input signal
$\tilde{m}(t)$ = reference signal

Output of the comparator is

$$d(t) = \Delta \times sgn[e(t)] = \begin{cases} \Delta & e(t) > 0 \\ -\Delta & e(t) < 0 \end{cases}$$

Thun the output of the delta modulator is

$$\begin{aligned} x_{DM}(t) &= \Delta \times sgn[e(t)] \textstyle\sum_{n=-\infty}^{\infty} \delta(t - nT_s) \\ &= \Delta \times \textstyle\sum_{n=-\infty}^{\infty} sgn\,[e(nT_s)]\delta(t - nT_s) \end{aligned}$$

Thus the output of the delta modulator is a series of impulse each having positive or negative polarity depending on the sign of $e(t)$ at the sampling instant. Integrating $x_{DM}(t)$, we obtain

$$\tilde{m}(t) = \textstyle\sum_{n=-\infty}^{\infty} \Delta \times sgn[e(nT_s)]$$

which is staircase approximation of m(t).

3.4.1 Demodulation:

A low pass filter, a summer, and a delay circuit are used to create a delta demodulator. This eliminates the prediction circuit, so the demodulator receives no assumed input.

Advantages of DM comparing with DPCM :

- 1-bit quantizer
- modulator and demodulator have a very simple design.

However, DM has some noise.

- Slope Over load distortion (when Δ is small)
- Granular noise (when Δ is large)

3.5 NOISE IN PCM

Channel noise and quantization noise (or quantization error) are the two different kinds of noise that affect the PCM.

3.5.1 Channel Noise

Different disturbances present on the channel anywhere in between the transmitter output and receiver input leads to channel noise to get introduced. The same is likely to be present once the communication equipment is turned on. The channel noise mainly causes bit errors in the received signal. In the presence of the channel noise, average probability of symbol error, which is defined as the probability that the reconstructed symbol at the receiver output differs from the binary symbol on an average, can be used to assess the accuracy of information conveyed with PCM. Channel noise can be eliminated by using regenerative repeater (RR). RR is basically a threshold comparator which compares the incoming signal with a certain threshold value and generates a new pulse.

3.5.2 Quantization Noise

The quantization noise is introduced by the quantizer. It is the difference between the message signal's original value and the quantizer's output. The same is expressed as,

$$m(t) - Q[m(t)] \tag{3.7}$$

where $Q[m(t)]$ is the quantized value of m(t). The maximum quantization error ($[q_e]_{max}$) is given by,

$$[q_e]_{max} = \pm\frac{\Delta v}{2} \tag{3.8}$$

where Δv is step size of the quantizer which is given as,

$$\Delta v = \frac{V_{max} - V_{min}}{2^n} = \frac{m_p - (-m_p)}{2^n} = \frac{2m_p}{2^n} \tag{3.9}$$

where $V_{max} = m_p$ and $V_{min} = -m_p$. Thus, the quantization error q_e lies uniformly anywhere in the range of $\left(-\frac{\Delta v}{2}, \frac{\Delta v}{2}\right)$. The probability distribution function ($f_{q_e}(q)$) of such a uniform random variable is shown in Fig. 3.18 and is expressed as,

$$f_{q_e}(q) = \begin{cases} 0; & q < -\frac{\Delta v}{2} \\ \frac{1}{\Delta v}; & -\frac{\Delta v}{2} \leq q \leq \frac{\Delta v}{2} \\ 0; & q > \frac{\Delta v}{2} \end{cases}$$

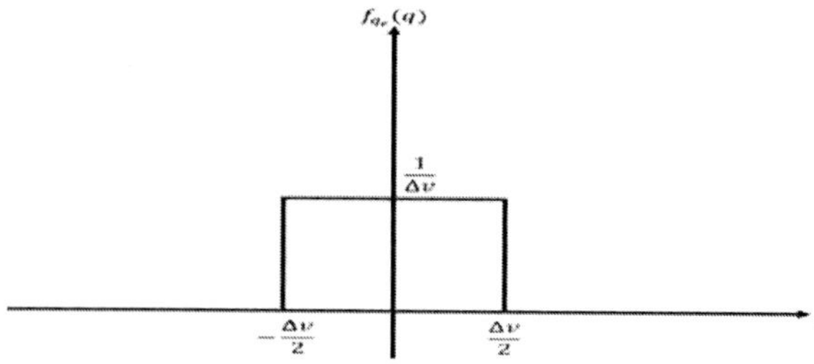

Fig 3.18 Probability distribution function

Now, as the mean value of any random variable X is expressed as:

$$E(\mathrm{X}) = \int_{-\infty}^{\infty} x f_x(x) dx. \tag{3.10}$$

Thus, the mean ($\langle q_e \rangle$) of the quantization error is given as,

$$\langle q_e \rangle = \int_{-\infty}^{\infty} q_e \frac{1}{\Delta v} dq_e = \int_{-\Delta v/2}^{\Delta v/2} q_e \frac{1}{\Delta v} dq_e = 0. \tag{3.11}$$

Similarly, variance of X is expressed as:

$$E(X^2) = \int_{-\infty}^{\infty} x^2 f_X(x) dx. \tag{3.12}$$

Then, the mean square quantization error ($\langle {q_e}^2 \rangle$) or power of quantization noise is given as

$$\langle {q_e}^2 \rangle = \int_{-\infty}^{\infty} {q_e}^2 \frac{1}{\Delta v} dq_e = \int_{-\Delta v/2}^{\Delta v/2} {q_e}^2 \frac{1}{\Delta v} dq_e \tag{3.13}$$

The mean square quantization error can be further expressed as,

$$\langle {q_e}^2 \rangle = \frac{1}{\Delta v} \left(\frac{{q_e}^3}{3} \right)_{-\Delta v/2}^{\Delta v/2} = \frac{1}{3\,\Delta v} \left(\frac{\Delta v^3}{8} + \frac{\Delta v^3}{8} \right) = \frac{\Delta v^3}{12\,\Delta v} = \frac{\Delta v^2}{12} \tag{3.14}$$

Now, consider a sinusoidal message signal $m(t) = A_m \sin \omega t$. The signal power (S) and noise power (N) are then given by,

$$N = \left(\frac{2m_p}{2^n} \right)^2 \times \frac{1}{12} = \frac{{m_p}^2}{3 \times 2^{2n}} \tag{3.15}$$

$$S = \frac{{m_p}^2}{2} \tag{3.16}$$

The SNR $\left(\frac{S}{N}\right)$ is then given as,

$$\frac{S}{N} = \frac{3}{2} \cdot 2^{2n} \approx (1.8 + 6n)\mathrm{dB} \tag{3.17}$$

Quantization error should be minimum for proper reconstruction of message signal. Hence, the step size should be reduced either by decreasing the dynamic range ($V_{max} - V_{min}$) or by increasing bit numbers per symbol. However, in a PCM system, the option to change the dynamic range is not available. Further, a higher value of n though reduces the noise, it also causes the bandwidth to increase. This is the one of the major drawbacks of PCM.

3.6 DIGITAL MULTIPLEXER

Multiplexing is the technique of integrating numerous signals into one signal through a common media; analog multiplexing is the process of multiplexing analog signals. Similarly, the multiplexing of digital signals is known as digital multiplexing. Multiplexing is a fundamental characteristic of all commercial long-distance communication systems. It relies on the interleaving of symbols from two or more digital signals. Digital multiplexing will allow us to combine several digital signals, including computer outputs, digital speech, digital fax, and television signals.

The digital data can be multiplexed by using a bit-by-bit interleaving procedure, it is done by using a selector switch which sequentially selects a bit from each input and places it over the high-speed transmission line. At receiving end, the bits received on the common line are separated out and delivered to their respective destinations.

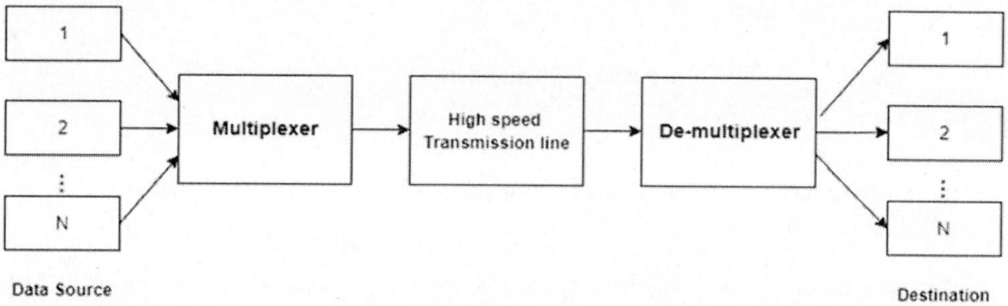

Fig 3.19 Digital multiplexing and de-multiplexing

Functions performed by multiplexers are:

1. To establish a frame, a frame consists of at least one bit from every input.
2. A number of unique bits slots within the frame should be assigned to each input.
3. We insert control bits for frame identification and synchronisation.
4. To make allowance for any variation of the input bit rates.

Types of Multiplexers:

1. Synchronous multiplexer
2. Asynchronous multiplexer
3. Quasi-synchronous multiplexer

3.6.1 Time Division Multiplexing

The Time Division Multiplexing (TDM) is a multiplexing technique in which each user is given the entire access to the whole bandwidth of the channel for a fixed duration of time. After this, the control is moved to the next user, and the process continues. In TDM multiple data streams as shown in Fig. (3.20) are put together in a single signal with the help of a multiplexer by separating the signal into many segments, each allotted a fixed duration of time. Each individual data streams thus transmitted is then reassembled at the receiver end based on their arrival time using a demultiplexer.

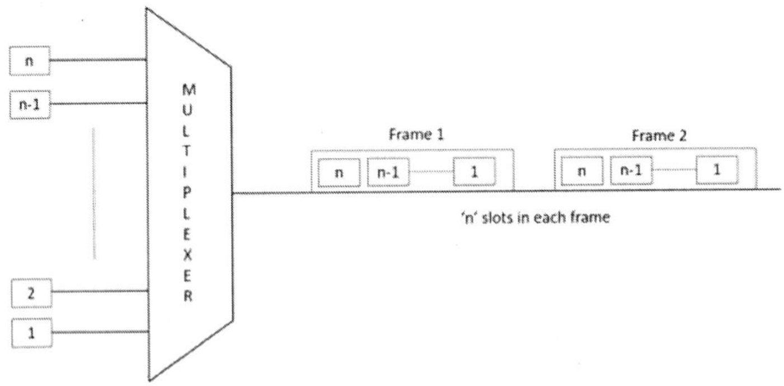

Fig 3.20 Time Division Multiplexer

In TDM, all the signals operate with the same frequency at different times. It is of following types:

1) Synchronous TDM: In this, time slots allocated to each user are pre-assigned and fixed. The slot is even given to the user which is not ready with data or has no data to transmit at the given time. In this case, the particular slot is transmitted empty.

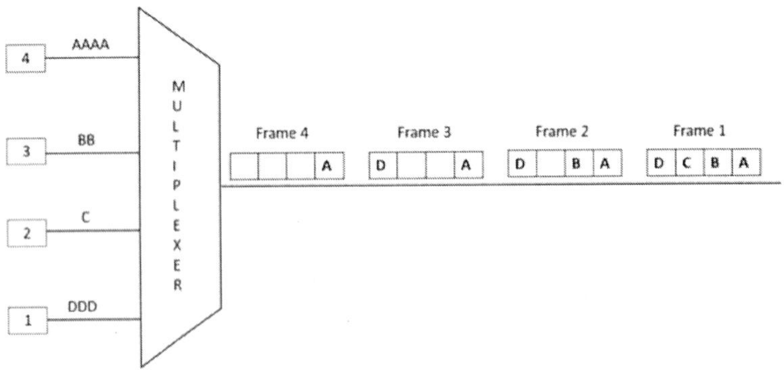

Fig 3.21 Synchronous TDM

1) Asynchronous TDM: In this, the time slots are allocated dynamically depending on the user's ready state or whether they have information to transmit in each time slot. It dynamically allocates the time slots according to different user's requirement. Asynchronous TDM provides more channel capacity compared to synchronous TDM.

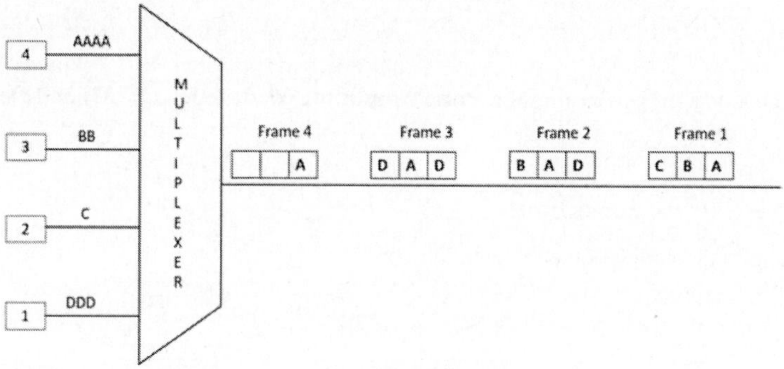

Fig 3.22 Asynchronous TDM

3.6.2 Frequency Division Multiplexing

The multiplexing technique known as frequency division multiplexing (FDM) combines numerous signals across a single carrier. Signals from many frequencies are merged for simultaneous transmission in FDM. FDM divides the total bandwidth into non-overlapping frequency bands. Each of these bands is a carrier for a different signal that is generated and modulated by one of the sending devices. To prevent signal overlapping, the frequency bands are separated by guard bands, which are strips of unused frequencies.

To combine the modulated signals, a multiplexer (MUX) is employed at the transmitting end. The combined signal is broadcast through the communication channel, enabling various data streams to be exchanged simultaneously. Demultiplexing removes individual signals from the combined signal at the receiving end (DEMUX), as we can see in Fig (3.23).

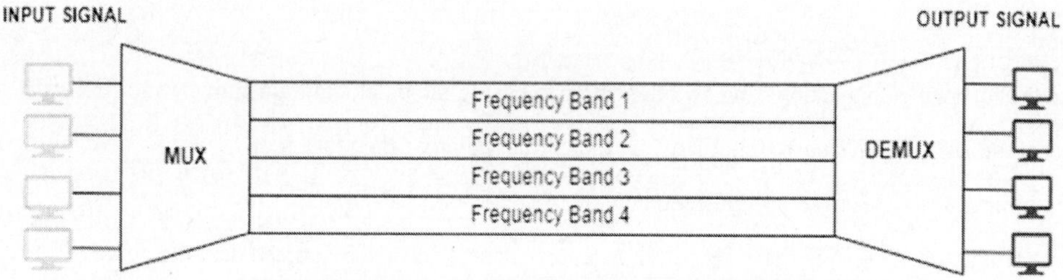

Fig 3.23 FDM-multiplexing and de-multiplexing

Uses and applications:

1) It allows multiple independent signals generated by multiple users to share a single transmission medium, such as a copper cable or a fibre optic cable.

2) FDM has been widely used in telephone networks to multiplex calls. It can also be used in satellite communications, cell phone networks, and wireless networks.

PRACTICAL

Experiment 1: To study the operation of a Pulse Amplitude Modulation (PAM) and Demodulation.

Objective:

Study of pulse amplitude modulation and demodulation.

Theory:

In pulse amplitude modulation, the signal is sampled at regular intervals, with each sample's amplitude being proportional to the signal's amplitude at the time of sampling. Each sample's amplitude is maintained for the duration of the sample to create flat-topped pulses. A Butterworth filter with an active lowpass is used as the pulse amplitude demodulator. It removes the sampling frequency and its harmonics from the modulated signal and then, through an integrated process, restores the baseband signal.

Equipment:

- Experimenter kit **DCL- 08**
- Connecting chords
- Power supply
- **20 MHz** dual trace oscilloscope

Note: Keep the switch FAULT IN off position

Procedure:

- Make the following connections and switch settings by using the block diagram shown in Fig 1.
- Connect the DCL-08 kit's power supply using the correct polarity, then turn it on.
- Using jumper JP1, select a sampling frequency of 16 KHz.
- Connect the PAM IN post to the 1 kHz, 2 V peak to peak sine wave signal created on the board.
- Check the PAM OUT post's pulse amplitude modulation output.
- Reduce the length of the ensuing posts by adding the connecting chords as shown in the block diagram for PAM OUT and AMP IN and AMP OUT and FIL IN.
- Continue turning the amplifier gain control potentiometer P5 all the way clockwise.

- Check that the demodulated pulse aptitude signal at FIL OUT is identical to the input signal.
- Repeat the test with various input signals and sampling rates.

Switch Faults:

- Put switch 1 of SF1 in Switch Fault section to ON position. The feedback resistor is bypassed from Amplifier section. Gain of Amplifier now depends on potentiometer P5 only.
- Put switch 2 of SF1 in Switch Fault section to ON position. This will generate two mixed sine waves, which could be used as a modulating input signal for modulators PAM, PWM and PPM.
- Put switch 3 of SF1 in Switch Fault section to ON position. This will bypass one filter from filter section. The output consists of ripple with reference to previous output without switch fault.
- Put switch 4 of SF1 in Switch Fault section to ON position. This provides constant high sampling signal to the sampling switch, which in turn gives natural sampling at the output.
- Put switch 5 of SF2 in Switch Fault section to ON position. This removes the control signal of first switch of PAM section, this will open pin of CMOS IC. Due to this output will be abrupt or may follow the input.

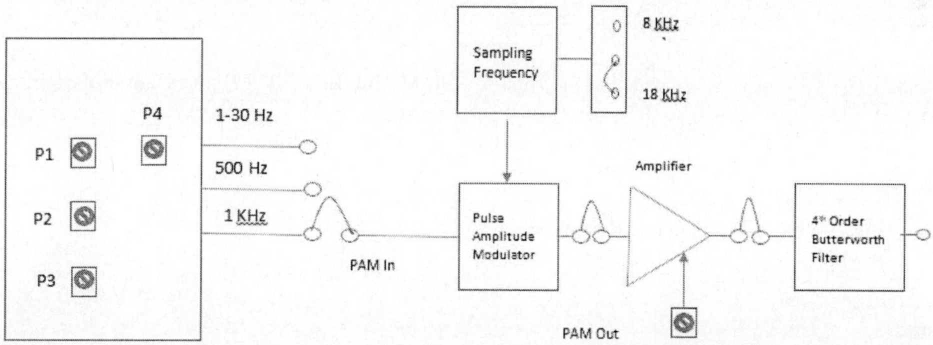

Fig 1: Block Diagram of Pulse Amplitude Modulation (PAM) and Demodulation

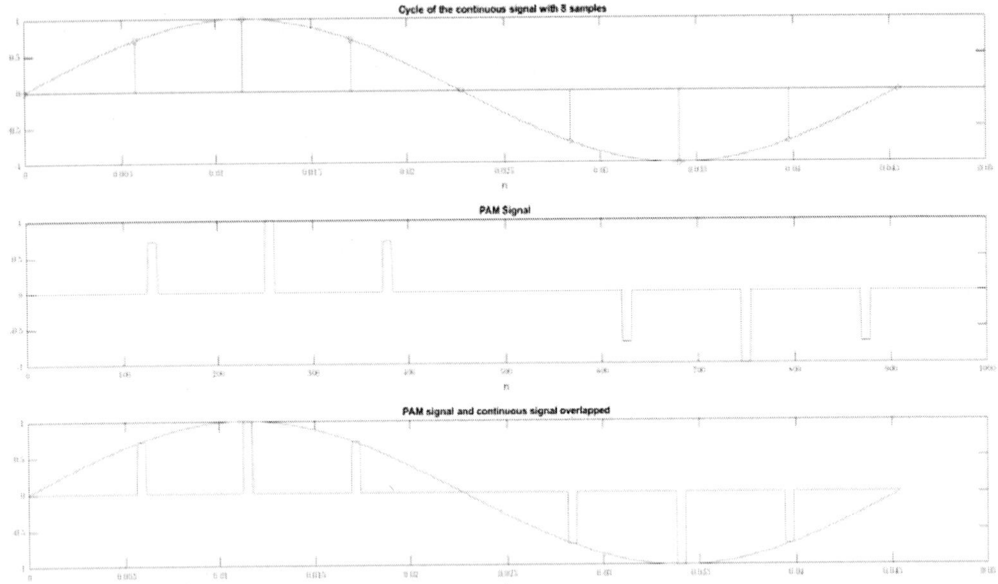

Fig 2: PAM Waveforms

Experiment 2: To study the operation of a Pulse Width Modulation (PWM) and Demodulation.

Objective:

Study of pulse width modulation and demodulation

Theory:

Pulse width modulation

This modulation method regulates the square wave's duty cycle fluctuation in response to the input modulating signal (which has a certain fundamental frequency). Here, the square wave's on-period variation reflects the modulation signal's amplitude variation. Consequently, it is a V to T conversion technique.

Pulse width demodulation

The input signal is pulse width modulated signal, so the ON time of signal is changing according to the modulating signal. One counter is activated in this demodulation method while the PWM signal is active. The counter outputs a certain count at the end of the ON period that exactly matches the input signal's amplitude. Then this count is supplied to a DAC. The amplitude of the input signal is reflected in the DAC's output. As a result, a train of varying pulse width produces varying count values, and a DAC responds by producing outputs that are directly proportional to the input signal's amplitude. The original signal is then obtained by filtering this. As a result, the PWM wave's original modulating signal is what we receive at the output.

Equipment:

- experimenter kit DCL 08
- Connecting cords
- Power supply
- 20-Megahertz dual trace oscilloscope

Note: Keep the switch fault in off position

Procedure:

- Make the following connections and switch settings while consulting block diagram Fig. 2.
- Connect the DCL-08 kit's power supply using the correct polarity, then turn it on.
- Position jumper J P3 in the second spot.
- Choose an onboard sine wave signal with a peak-to-peak voltage of 1 kHz.
- Attach this signal to the PWM/PPM IN port.
- Pay attention to the PWM OUT post's pulse width modulated output. Because of the persistence of eyesight, only a blurred band in the waveform will be visible due to the high sampling frequency. Absent an input signal, all that can be seen is a square wave with a fundamental frequency that is fixed in time and no breadth change. Use a 1- 30 Hz sine wave signal applied to the PWM/PPM IN post to monitor the variation in pulse width. Change the frequency between 1 and 30 Hz.
- Short the following posts with the connecting chords provided as shown in block diagram for demodulation section.

PWM OUT and **BUF IN**
BUF OUT and **PWM DMOD IN**
DMOD OUT and **FIL IN**

- Check the output of the pulse width demodulator at FIL OUT.
- Use jumper JP3 to repeat the experiment with different input signals and sampling clocks.

Note: Procedure for observation of PWM output in DUAL mode

- Keep CH1 knob CRO on 1 Volt/div ac.
- Keep CH2 knob CRO on 2 Volt/div ac.
- Keep time per division knob on one millisecond
- Keep the CRO in dual channel mode. Use X10 for expansion
- After proper triggering of CRO, observe both the signal **PWM IN** and **PWM OUT** simultaneously.

Switch Faults:

- Put switch 2 of SF1 in Switch Fault section to ON position. This will generate two mixed sine waves, which could be used as a modulating input signal for modulators PAM, PWM and PPM.
- Put switch 3 of SF1 in Switch Fault section to ON position. This will bypass one filter from filter section. The output consists of ripple with reference to previous output without switch fault.
- Put switch 6 of SF2 in Switch Fault section to ON position. This opens the connection between PPM Demodulation & Latch enable signal pin of Latch at PPM/PWM demodulator. Demodulated PPM/PWM output is not obtained.
- Put switch 7 of SF2 in Switch Fault section to ON position. This will open MSB from the 8-bit input to DAC at Demodulator section of PWM. Output level reduces in amplitude.
- Put switch 8 of SF2 in Switch Fault section to ON position. This will change Minimum Pulse Width to zero level. Due to this PVM output will be absent for no signal at the PWM input. Minimum pulse width at the output of PWM modulator is not maintained Pulse width will very according to modulating signal.

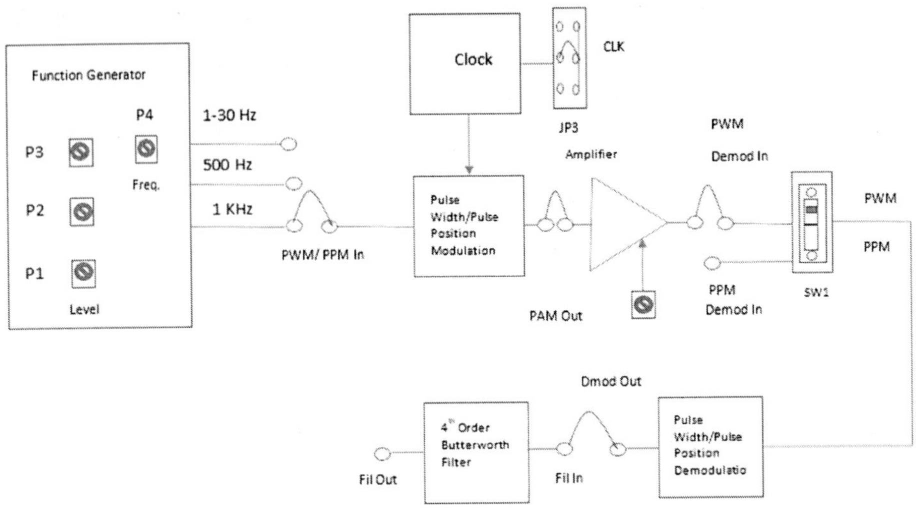

Fig 1: Block diagram of Pulse Width Modulation (PWM) and Demodulation.

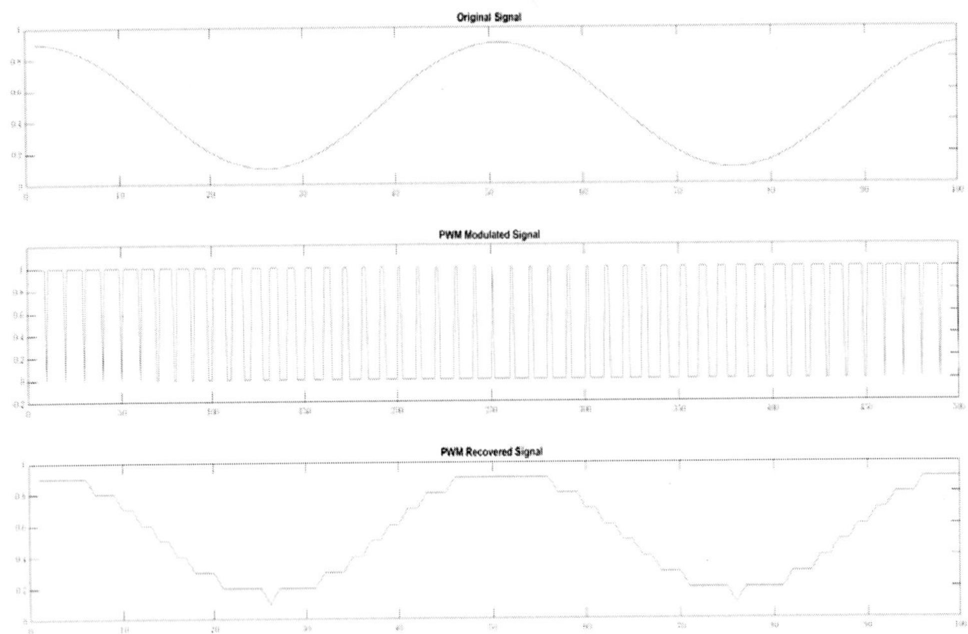

Fig 2: PWM Waveforms

Experiment 3: To study the operation of a Pulse Position Modulation (PPM) and Demodulation.

Objective:

Study of pulse position modulation and demodulation

Theory:

When the input modulating signal's amplitude varies, the TTL pulse's position changes on a time scale, but the pulse's width and amplitude remain constant.

Demodulation:-

Using a monostable multivibrator, the pulse position modulated signal is converted into PWM pulse form. The same PWM demodulation method is then used to demodulate this signal. One counter is enabled in this demodulation technique during the ON period of the PWM signal. The counter outputs a certain count at the completion of the ON period that is directly related to the input signal's amplitude. Then this count is supplied to a DAC. The output of the DAC matches the input signal's amplitude. As a result, a train of pulses with different pulse widths produces different count values, and the DAC responds by producing outputs that are directly proportional to the input signal's amplitude. The original signal is then obtained by filtering this. As a result, the PWM wave's original modulating signal is obtained at the output.

Equipment:

- Experimenter kit DCL 08
- Connecting chords
- Power supply
- 20 MHz dual trace oscilloscope

Note: Keep the switch faults in off position

Procedure:

- Use Fig. 3 as a reference for the block diagram and make the following connections and switch settings.
- Connect the DCL-08 kit's power supply using the correct polarity, then turn it on.
- Position jumper JP3 in the second spot.
- Choose the onboard generated 1 kHz 1 VPP sine wave signal.
- Connect the specified signal to PWM/PPM IN.
- Observe the pulse position modulated output at PPM OUT post with shifting location on time scale.
- Please take note that the pulse's width and amplitude are identical, and that the shift in the pulse's position is proportional to the input analogue signal. Apply a sine wave signal with a frequency range of 1–30 Hz to the PWM/PPM IN post and observe the signal in a dual oscilloscope for the posts PPM OUT and PWM OUT at the same time to observe the variation in pulse position.Then short the following posts with the link provided as shown in block diagram for demodulation section.

PPM OUT and **BUF IN**
BUF OUT and **PPM DMOD IN**
DMOD OUT and **FIL IN**

- Keep the pulse position demodulated signal at FIL OUT while observing.

• Perform the experiment again with a different input signal and sample frequency.

Switch Faults:

- Put switch 2 of SF1 in Switch Fault section to ON position. This will generate two mixed sine waves, which could be used as a modulating input signal for modulators PAM, PWM and PPM.
- Put switch 3 of SF1 in Switch Fault section to ON position. This will bypass one filter from filter section. The output consists of ripple with reference to previous output without switch fault.
- Put switch 6 of SF2 in Switch Fault section to ON position. This opens the connection between PPM Demodulation & Latch enable signal pin of Latch at PPM/PWM demodulator. Demodulated PPM/PWM output is not obtained.
- Put switch 7 of SF2 in Switch Fault section to ON position. This will open MSB from the 8-bit input to DAC at Demodulator section of PWM. Output level reduces in amplitude.
- Put switch 8 of SF2 in Switch Fault section to ON position. This will change Minimum Pulse Width to zero level. Due to this PWM output will be absent for no signal at the PWM input. Minimum pulse width at the output of PWM modulator is not maintained Pulse width will very according to modulating signal.

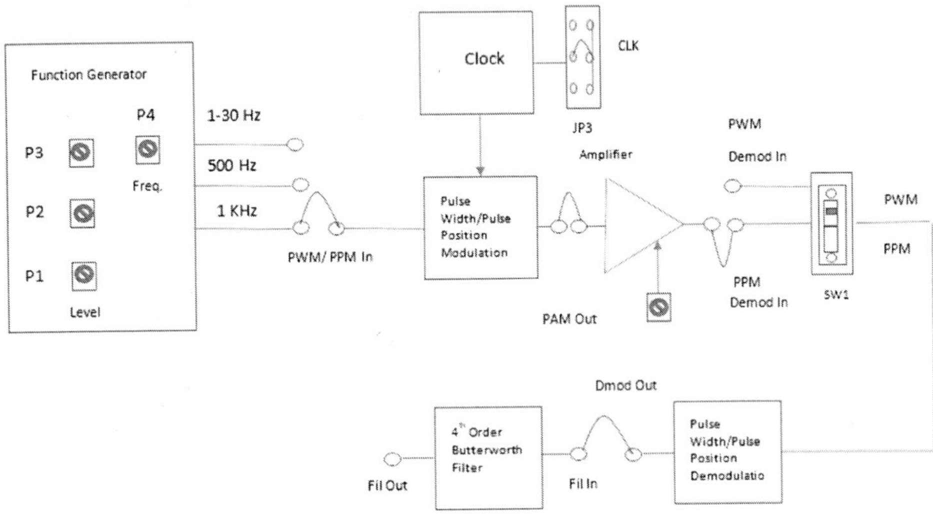

Fig 1: Block Diagram of Pulse Position Modulation (PPM) and Demodulation

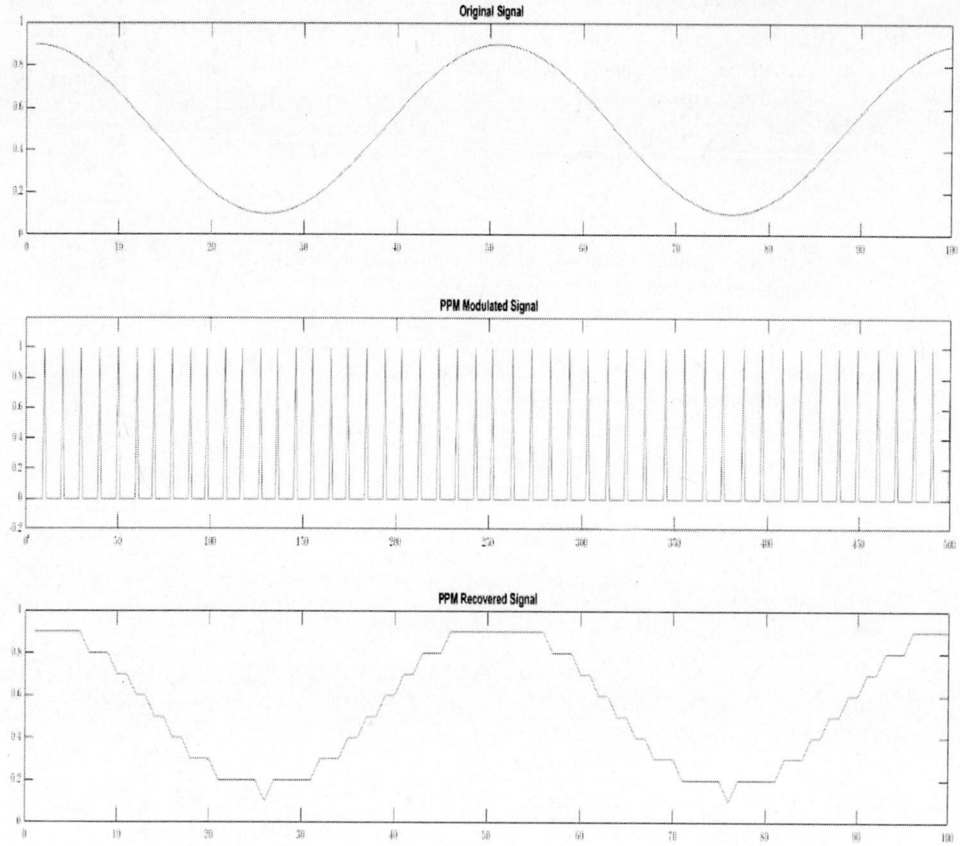

Fig 2: PPM Waveforms

References and Suggested Reading

1. B. P. Lathi, and Zhi Ding, "Modern Digital and Analog Communication Systems", Oxford University Press, 2010.

2. Simon Haykin, "Communication Systems", Wiley, 2007.

3. Herbert Taub, Donald L. Schilling, and Gautam Saha, "Principal

of Communication Systems", McGraw Hill Education, 2017.

4. P.F. Panter, "Modulation, Noise and Spectral Analysis", McGraw-Hill, New York, 1965.
5. https://nptel.ac.in/courses/117102059
6. https://ocw.mit.edu/courses/16-36-communication-systems-engineering-spring-2009/pages/lecture-notes/
7. George Clifford Hartley, "Techniques of Pulse-code Modulation in Communication Networks", 1967.

4 BASEBAND RECEIVER

UNIT SPECIFICS

Through this unit we have discussed the following aspects:

- *Elements of detection theory*
- *Optimum detection of signal in noise,*
- *Probability of error evaluations,*
- *Baseband Pulse transmission,*
- *Intersymbol interference and Nyquist criterion,*

The practical applications of the topics are discussed for generating further curiosity and creativity as well as improving problem solving capacity.

Besides giving a large number of multiple choice questions as well as questions of short and long answer types marked in two categories following lower and higher order of Bloom's taxonomy, assignments through a number of numerical problems, a list of references and suggested readings are given in the unit so that one can go through them for practice. It is important to note that for getting more information on various topics of interest some QR codes have been provided in different sections which can be scanned for relevant supportive knowledge.

After the related practical, based on the content, there is a "Know More" section. This section has been carefully designed so that the supplementary information provided in this part becomes beneficial for the users of the book. This section mainly highlights the initial activity, examples of some interesting facts, analogy, history of the development of the subject focusing the salient observations and finding, timelines starting from the development of the concerned topics up to the recent time, applications of the subject matter for our day-to-day real life or/and industrial applications on variety of aspects, case study related to environmental, sustainability, social and ethical issues whichever applicable, and finally inquisitiveness and curiosity topics of the unit.

RATIONALE

This unit on baseband receiver helps students to get a primary idea about the optimum detection of signa. We have first started with the elements of estimation detection theory, where we have introduced basic of detection. Then, we have covered probability of error evaluation, baseband pulse transmission and intersymbol interference and Nyquist criterion in a detailed manner.

Pulse modulation is a type of analogue modulation technique which is employed on discrete information signals. The continuous signals which are obtained from nature can be converted to discrete signals with the help of sampling as is explained in this chapter.at the end of this chapter we will understand the advantages of pulse modulation, various effects which deteriorate the performance and then later on as we progress we will learn about the ways to mitigate them.

PRE-REQUISITES

Mathematics: Calculus (Class XII)

Physics: Basic of Communication (Class XII)

UNIT OUTCOMES

List of outcomes of this unit is as follows:

U1-O1: Study of baseband receivers optimum detection of signal in noise.

U1-O2: Explain probability of error evaluations.

U1-O3: Describe baseband pulse transmission.

U1-O4: Explain Intersymbol interference and Nyquist criterion,

This table below need to be filled after discussion as marked with red.

Unit-1 Outcomes	EXPECTED MAPPING WITH COURSE OUTCOMES (1- Weak Correlation; 2- Medium correlation; 3- Strong Correlation)					
	CO-1	CO-2	CO-3	CO-4	CO-5	CO-6
U1-O1	3	3	3	-	3	1
U1-O2	1	1	2	2	1	-
U1-O3	2	1	3	1	2	1
U1-O4	-	-	3	1	2	2
U1-O5	3	3	3	-	3	1

4.1 ELEMENTS OF DETECTION THEORY

Using evidence to draw conclusions is the focus of detection theory. The presumed model which may have generated the data is the basis for decisions. Making decisions includes examining the data and identifying the model that is most likely to have generated them. By doing this, we can determine whether model was accurate. Signal processing is rife with decision issues. Determining whether the most recent bit received in the context of channel disruptions was a one or a zero, for instance, is a detection challenge in digital communications.

More specifically, we indicate by $\mathcal{M}_i$ the ith model that may have produced the data R. The conditional probability distribution of the data, given by the vector R, represents a "model". For instance, model i is represented by $p_{R|\mathcal{M}i}(r)$. Selecting the model that most closely matched what's been observed requires considering all possible models that could explain the data.

4.1.1 Likelihood Ratio Test

There are four potential outcomes for a binary detection problem with two models. The observation range is divided into two distinct decision zones, Z_0 and Z_1, which are used to make the decision. R has only two possible values: Z_0 or Z_1. We will state that "model $\mathcal{M}_0$ was true"; if a given R is in Z_0; otherwise, "model $\mathcal{M}_1$ was true" would be declared to be true.

The Bayes' decision criterion aims to reduce a decision-making cost function. Let C_{ij} represent the cost of choosing the incorrect model $(i \neq j)$ and C_{ii}represent the apparently lower cost of selecting the correct model (i.e., $C_{ij}$is more than C_{ii}, $i \neq j$). Let π_i represent model i's a priori probability.

Minimization of cost function gives us the condition (likelihood ratio test):

$$\frac{p_{R|\mathcal{M}1}(r)}{p_{R|\mathcal{M}0}(r)} \underset{\mathcal{M}_0}{\overset{\mathcal{M}_1}{\gtrless}} \frac{\pi_0(C_{10}-C_{00})}{\pi_1(C_{01}-C_{11})} \tag{4.1}$$

The likelihood ratio $\frac{p_{R|\mathcal{M}1}(r)}{p_{R|\mathcal{M}0}(r)}$,symbolically represented by $\Lambda(r)$. The optimal decision rule then evaluates this scalar-valued outcome against a threshold where η equals $\frac{\pi_0(C_{10}-C_{00})}{\pi_1(C_{00}-C_{11})}$. The likelihood ratio test is just a comparison between the likelihood ratio and a threshold.

$$\Lambda(r) \underset{\mathcal{M}_0}{\overset{\mathcal{M}_1}{\gtrless}} \eta \tag{4.2}$$

However, using a logarithm (monotonic transformation) on both sides, often known as the log-likelihood, can sometimes simplify the calculations required for the likelihood ratio test. We explicitly define the likelihood ratio test as

$$\ln(\Lambda(r)) \underset{\mathcal{M}_0}{\overset{\mathcal{M}_1}{\gtrless}} \ln(\eta) \tag{4.3}$$

4.1.2 Maximizing the Probability of a Correct Decision

Given that each set of data can only be described by one model (the models being mutually exclusive), the probability for correctly distinguishing the two models, P_c is given by

$$P_c = \Pr[\ \mathcal{M}_0 \text{ , when } \mathcal{M}_0 \text{ is true}] + \Pr[\ \mathcal{M}_1\text{, when } \mathcal{M}_1 \text{ is true}] \tag{4.4}$$

When we translate the probability of correct decision in terms of likelihood functions $p_{R|\mathcal{M}i}(r)$, a priori probabilities, and the decision regions, we arrive to the following expression:

$$P_c = \int \pi_0\, p_{R|\mathcal{M}0}(r)dr + \int \pi_1\, p_{R|\mathcal{M}1}(r)dr \tag{4.5}$$

By choosing the Z_0 and Z_1 decision regions, we aim to maximize P_c. Select $\mathcal{M}_i$ for which the product $\pi_i\, p_{R|\mathcal{M}i}(r)$ is maximum for given r. When choosing between two models, simple manipulations result toward the likelihood ratio test.

$$\frac{p_{R|\mathcal{M}1}(r)}{p_{R|\mathcal{M}0}(r)} \underset{\mathcal{M}_0}{\overset{\mathcal{M}_1}{\gtrless}} \frac{\pi_0}{\pi_1} \tag{4.6}$$

Instead of the probability of being accurate, we typically compute the probability of error P_e, to estimate the quality of the decision rule. The observations, likelihood ratio, and sufficient statistic can all be used to express this quantity.

$$\begin{aligned} P_e &= \pi_0 \int p_{R|M_0}(r)dr + \pi_1 \int p_{R|M_1}(r)dr \\ &= \pi_0 \int p_{\Lambda|\mathcal{M}_0}(\Lambda)d\Lambda + \pi_1 \int p_{\Lambda|\mathcal{M}_1}(\Lambda)d\Lambda \\ &= \pi_0 \int p_{\Upsilon|\mathcal{M}_0}(\Upsilon)d\Upsilon + \pi_1 \int p_{\Upsilon|M_1}(\Upsilon)d\Upsilon \end{aligned} \tag{4.7}$$

4.1.3 Neyman-Pearson Criterion

The a priori probabilities are explicitly used in the Bayesian decision rule. There are several circumstances where it is unreasonable to assign or measure the a priori probabilities π_i. Neyman-Pearson criterion can be used in these circumstances to construct the decision rule.

Detection Probability: $P_D = \Pr\,[\text{say } \mathcal{M}_1 \mid \mathcal{M}_1 \text{ true}]$
False-alarm Probability: $P_F = \Pr\,[\text{say } \mathcal{M}_1 \mid \mathcal{M}_0 \text{ true}]$
Miss Probability: $P_M = \Pr\,[\text{say } \mathcal{M}_0 \mid \mathcal{M}_1 \text{ true}]$

The relationship between detection and miss probability is $P_M = 1 - P_D$. These probabilities do not rely on the a priori probabilities because they are conditional probabilities. In addition, the errors when using any decision rule are described by the two probabilities P_F and P_D. When using the Neyman-Pearson criterion, we maximize the detection probability P_D while assuming that the false-alarm probability is limited to be less than or equal to a given value α.

4.2 OPTIMUM RECEIVER

An optimal receiver is one that is made to reduce the likelihood that a decision error would occur. It is of two types-

1. Correlation receiver
2. Matched Filter receiver

4.2.1 The Correlation Receiver

The correlation receiver was shown to be the optimum receiver for baseband signals. The signal received across the communication channel is compared by the correlation receiver to the collection of known reference signals. The mutual correlation coefficient R(t) of the signal received was calculated for the comparison.

The development of the optimal receiver for coherent bandpass signals now follows a similar path and results in the correlation receiver. The optimum threshold and probability of bit error P_b for symmetrical and asymmetrical bandpass signals is found to be the same as that for baseband signals.

Demodulation of bandpass digital communication systems can be accomplished either coherently, where the receiver internally produces a reference signal of the same frequency and phase as the input signal, or noncoherently, where the receiver does not use a reference signal, but employs either differential encoding of the information or even simpler means.

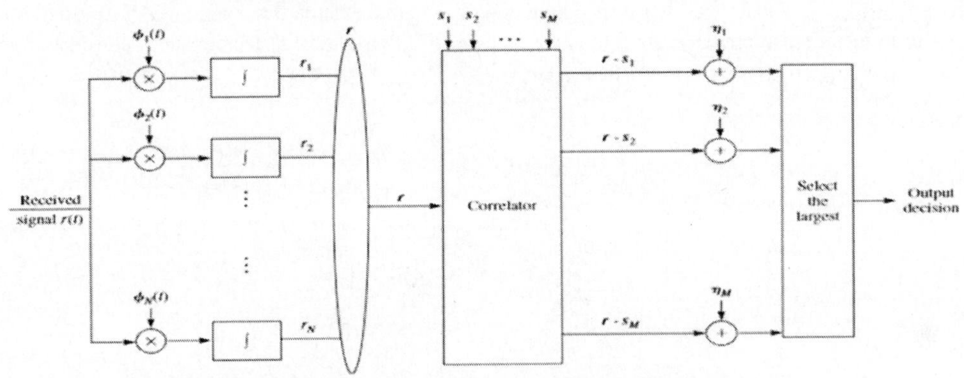

Fig. 4.1 The structure of a correlation receiver with N correlator

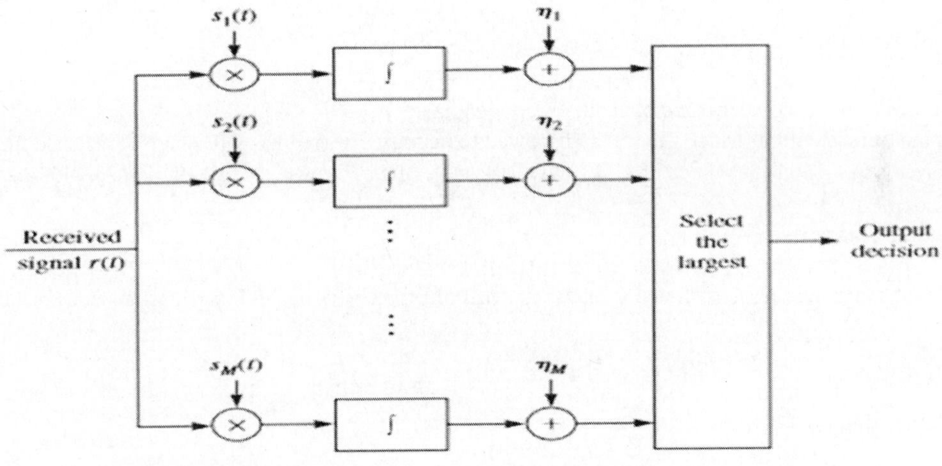

Fig. 4.2 The structure of the correlation receiver with M correlators

Although the structure in Fig. (4.2) appears to be simpler than the structure in Fig. (4.1), the correlation receiver of Fig. (4.1) is typically the chosen implementation strategy since in most circumstances $N < M$ (and in fact $N \ll M$). The correlation receiver needs N or M correlators, or multipliers followed by integrators.

The MAP decision rule is implemented by an optimal receiver for the AWGN channel provided by Equation below

$$\hat{m} = arg \max_{1 \le m \le M} [\eta_m + r.s_m], \ where \ \eta_m = \frac{N_0}{2} \ln P_m - \frac{1}{2} \varepsilon_m \quad (4.8)$$

Application- To measure spatial coherence function at given spatial separation on waveform, sampled by any two antennas. It may be useful in calculating wind probability distribution, sales volume per week etc.

4.2.2 Matched Filter Receiver

A filter is referred to as a matched filter if its output is produced in a fashion that maximizes the ratio of output peak power to mean noise power in its frequency response. This is a crucial factor that is considered while constructing any radar receiver.

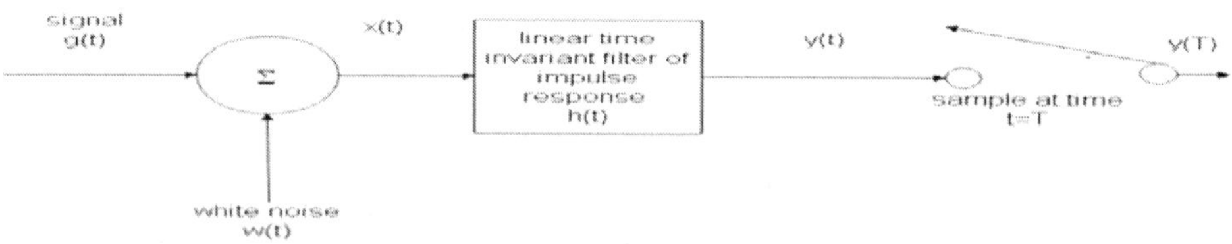

Fig. 4.3 Block Diagram matched filtering

Frequency Response: -

The complex conjugate of the input signal spectrum will determine how responsive the matching filter's frequency response is.

$$H(f) = S^*(f)\, e^{-j2\pi fT} \quad (4.9)$$

where $S(f)$ is the Fourier transform of the input signal s(t)
T is the time instant at which the signal observed to be max. The frequency response of matched filter receiver is having the magnitude of $S^*(f)$ and phase angle of $e^{-j2\pi fT}$ which varies uniformly with frequency.

Impulse Response: -

By using the inverse response function on the matched filter receiver's output in the time domain, we will obtain h(t).

$$h(t) = \int_{-\infty}^{\infty} H(f)\, e^{-j2\pi fT}\, df \quad (4.10)$$

and SNR output is $\frac{2\epsilon_s}{N_0}$.

Advantage: This filter enhances the signal to noise ratio by decreasing the spectral bandwidth wavelet through wavelet spectrum shape.

Disadvantage:

1. It needs past information of the primary user.
2. consumes more power.

Application:
This filter is used in a widely range of application that involves pattern recognition like detection of an image radar signal interpretation and FSK demodulation.

4.3 COHERENT COMMUNICATION

The requirements for Baseband Transmission is to provide -

A) Maximum data rate for a given Bandwidth.
B) Minimum probability of error.
C) Minimum transmitted power
D) Minimum Inter Symbol Interference.
E) Minimum circuit complexity.

4.3.1 Binary Amplitude Shift Keying (BASK) - It is a type of Amplitude Modulation which represents the binary data in the form of variations in the amplitude of a signal. Amplitude changes according to data stream.

Representation of BASK -

$$b(t) = \text{Binary '1'}: \; s_1(t) = A_c cos(\omega_c t) \qquad (4.13)$$
$$\text{Binary '0'}: \; s_2(t) = 0$$

$$s(t) = b(t).\, A_c cos(\omega_c t) \qquad (4.12)$$

BASK Waveform -

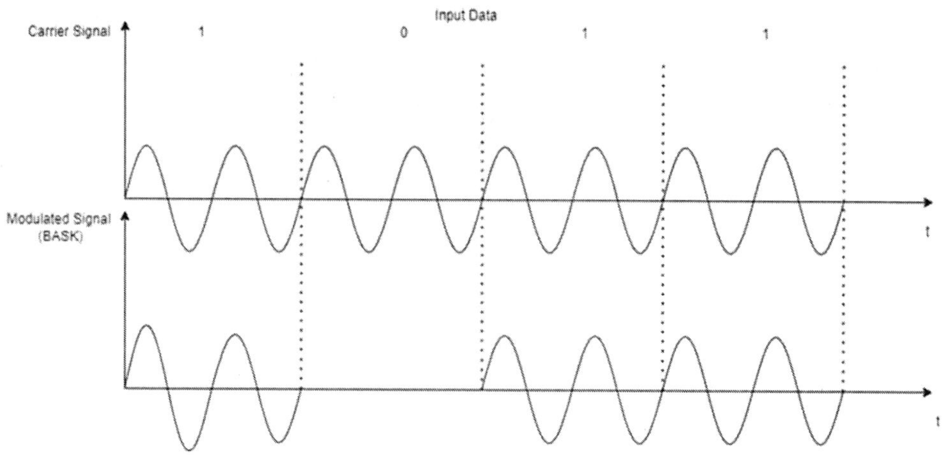

Fig. 4.4 BASK waveform

4.3.2 Frequency Shift Keying (FSK) - FSK is a digital modulation technique in which the frequency of the carrier signal varies according to the discrete digital changes. FSK is a scheme of frequency modulation. Frequency changes according to data stream.

Representation of BFSK -

$$b(t) = \text{Binary '1'} : s_1(t) = A_c cos2\pi(f_c + \Omega)t = A_c cos(2\pi f_H t) \quad (4.13)$$

$$\text{Binary '0'} : s_2(t) = A_c cos2\pi(f_c - \Omega)t = A_c cos(2\pi f_L t)$$

$$s(t) = b(t).\, A_c cos2\pi(f_c + b(t)\Omega)t \quad (4.14)$$

BFSK Waveform -

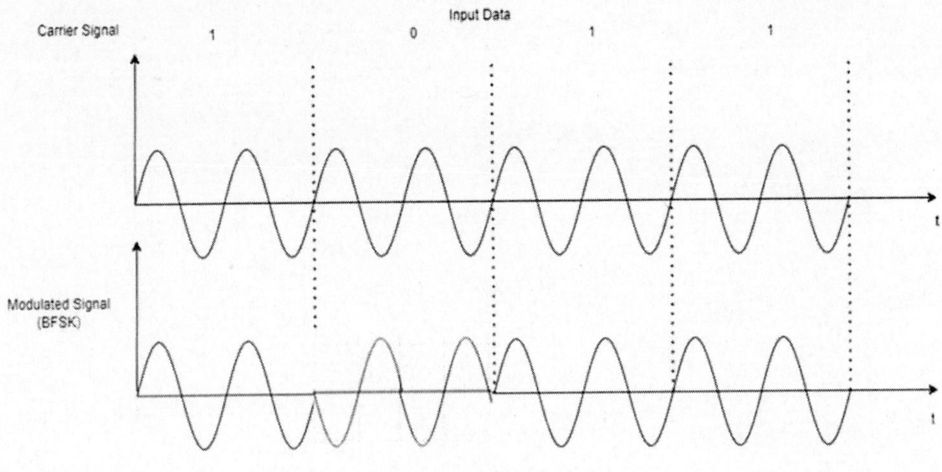

Fig. 4.5 BFSK waveform

4.3.3 Binary Phase Shift Keying (BPSK) - Phase Shift Keying (PSK) is the digital modulation technique in which the phase of the carrier signal is changed by varying the sine and cosine inputs at a particular time. Phase changes according to data stream.

Representation of BPSK -

b(t) = Binary '1' : $s_1(t) = A_c cos(\omega_c t)$ (4.15)

Binary '0' : $s_2(t) = A_c cos(\omega_c t + \pi)$

s(t) = b(t) . $A_c cos 2\pi(f_c + b(t)\Omega)t$ (4.16)

BPSK Waveform -

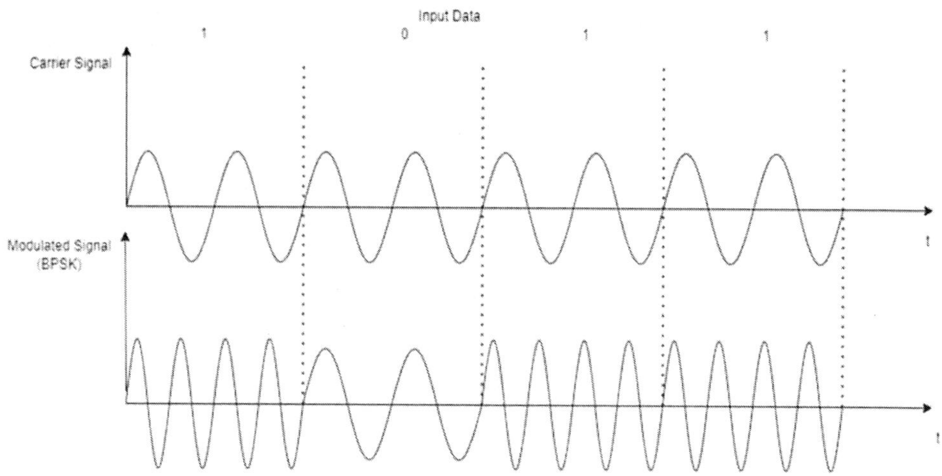

Fig. 4.6 BPSK waveform

4.4 PROBABILITY OF ERROR

$$P_e = Q\sqrt{\frac{E_d}{2\eta}}; \tag{4.17}$$

where $Q(x)$ is Q function, if x increases $Q(x)$ decreases, defined by

$$Q(x) = \int_x^{\infty} \frac{1}{\sqrt{2\pi}} \exp\left(\frac{-z^2}{2}\right).dz$$

where A_C is amplitude of carrier signal.
where η is noise power spectral density.
where E_d is difference energy signal, defined by

$$E_d = \int_0^{T_b} (s_1(t) - s_2(t))^2$$

4.4.1 BASK :

Binary '1' : $s_1(t) = A_c cos(\omega_c t)$
Binary '0' : $s_2(t) = 0$
$s_d(t) = s_1(t) - s_2(t) = A_c cos(\omega_c t)$
$E_d = \int_{-\infty}^{\infty} |s_d(t)|^2.dt = \int_0^{T_b} A_c^2 cos^2(\omega_c t).dt = \frac{A_c^2 T_b}{2}$

Therefore, $P_e = Q\sqrt{\frac{\frac{A_c^2 T_b}{2}}{2\eta}} = Q\sqrt{\frac{A_c^2 T_b}{4\eta}}$ (4.18)

- In ASK probability of error (Pe) is high and SNR is less.
- It has lowest noise immunity against noise.
- ASK is a bandwidth efficient system but it has lower power efficiency.

4.4.2 BFSK :

Binary '1' : $s_1(t) = A_c cos(2\pi f_H t)$
Binary '0' : $s_2(t) = A_c cos(2\pi f_L t)$

$s_d(t) = s_1(t) - s_2(t) = A_c cos(2\pi f_H t) - A_c cos(2\pi f_L t)$

$E_d = \int_{-\infty}^{\infty} |s_d(t)|^2.dt = \int_0^{T_b} |A_c cos(2\pi f_H t) - A_c cos(2\pi f_L t)|^2.dt$

$= \int_0^{T_b} {A_c}^2 cos^2(2\pi f_H t)dt +$
$\int_0^{T_b} {A_c}^2 cos^2(2\pi f_L t)dt - 2{A_c}^2 \int_0^{T_b} cos(2\pi f_H t) cos(2\pi f_L t)dt$

Since, $s_1(t)$ & $s_2(t)$ are orthogonal

$E_d = \frac{{A_c}^2 T_b}{2} + \frac{{A_c}^2 T_b}{2} = {A_c}^2 T_b$

$P_e = Q\sqrt{\frac{E_d}{2\eta}} = Q\sqrt{\frac{{A_c}^2 T_b}{2\eta}}$ (4.19)

- In case of FSK, Pe is less, and SNR is high.
- This technique is widely employed in modem design and development.
- It has increased immunity to noise but requires larger bandwidth compare to other modulation types.

4.4.3 BPSK :

Binary '1' : $s_1(t) = A_c cos(\omega_c t)$
Binary '0' : $s_2(t) = A_c cos(\omega_c t + \pi) = - A_c cos(\omega_c t)$

$s_d(t) = s_1(t) - s_2(t) = 2 A_c cos(\omega_c t)$

$E_d = \int_{-\infty}^{\infty} |s_d(t)|^2.dt = \int_0^{T_b} |2 A_c cos(\omega_c t)|^2 .dt =$
$4{A_c}^2 \int_0^{T_b} \frac{(1+cos(\omega_c t))}{2}.dt$

$E_d = 2{A_c}^2 T_b$

(4.40)

$$P_e = Q\sqrt{\frac{E_d}{2\eta}} = Q\sqrt{\frac{A_c{}^2T_b}{\eta}}$$

- In case of PSK probability of error is less. SNR is high.
- It is a power efficient system, but it has lower bandwidth efficiency.
- PSK modulation is widely used in wireless transmission.
- The variants of basic PSK and ASK modulations are QAM, 16-QAM, 64-QAM and so on.

4.4.4 Comparison Between BASK, BFSK, and BPSK -

Criteria	BASK	BFSK	BPSK
Probability of Error	$P_e = Q\sqrt{\frac{\frac{A_c^2T_b}{2}}{2\eta}} = Q\sqrt{\frac{A_c^2T_b}{4\eta}}$	$P_e = Q\sqrt{\frac{E_d}{2\eta}} = Q\sqrt{\frac{A_c{}^2T_b}{2\eta}}$	$P_e = Q\sqrt{\frac{E_d}{2\eta}} = Q\sqrt{\frac{A_c{}^2T_b}{\eta}}$
Constellation Diagram	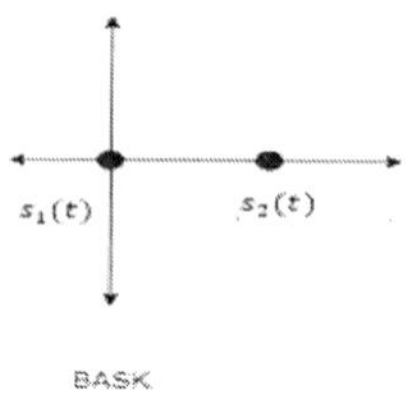	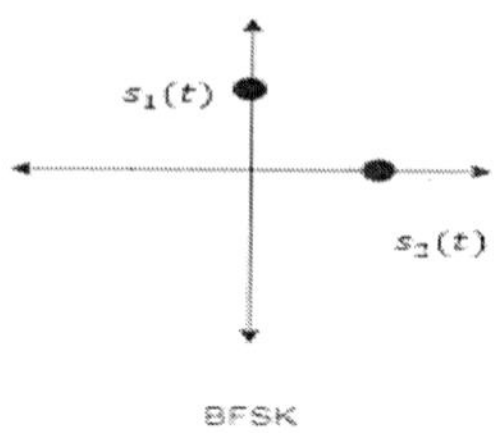	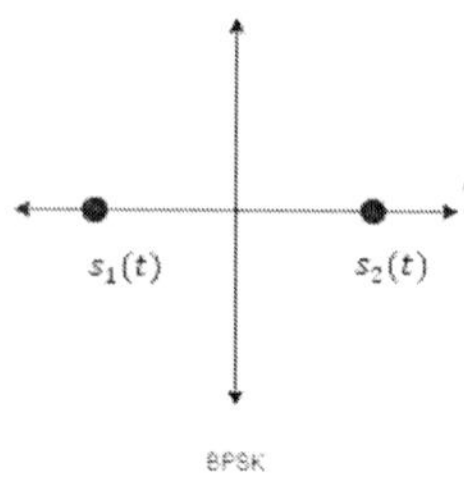
Signal to Noise Ratio	In ASK probability of error (Pe) is high and SNR is less.	In case of FSK, Pe is less and SNR is high.	In case of PSK probability of error is less. SNR is high.
Bandwidth	Less	More	Less to Moderate
Noise Immunity	Poor	Better	Better
Information	In amplitude variations	In frequency variations	In phase variations

4.5 SAMPLING

Sampling is the process of transforming an analog domain signal into a form compatible with a digital communication system. The sampling process links the analog waveform and its sampled version. The sampling process is classified into three types as given below

1. Ideal Sampling

Ideal sampling is also known as instantaneous or impulse train sampling. In this method, the sampling signal is a periodic impulse train. The sampled signal is obtained by multiplication of unit impulse train with the analog message signal. The area of each impulse in the sampled signal is equal to the instantaneous value of the input signal as shown in Fig.4.8. Ideal sampling can be achieved with the help of circuit shown in Fig. (4.7).

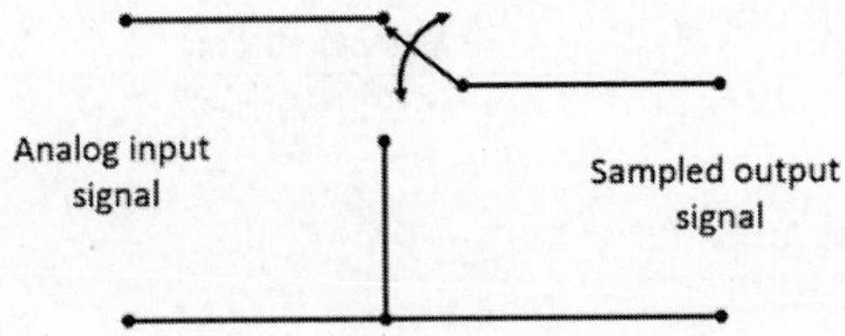

Fig. 4.7 Generation of ideal sampled signal

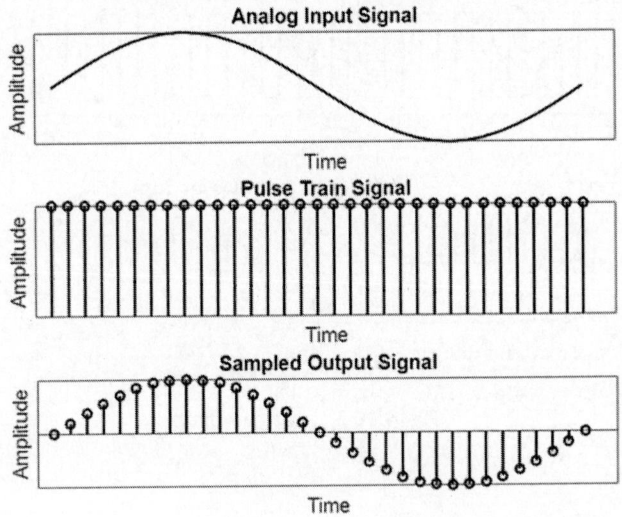

Fig. 4.8 Ideal signal sampling

2. Flat-top Sampling

Flat-top sampling is the simplest and the most popular sampling techniques. In this technique, the sampled pulse has the same amplitude throughout the sample duration. The amplitude of the analog signal at time instance t_0 i.e., the beginning of the sample duration is recorded and the same is maintained for the whole pulse hence the top of the pulses are flat in shape.

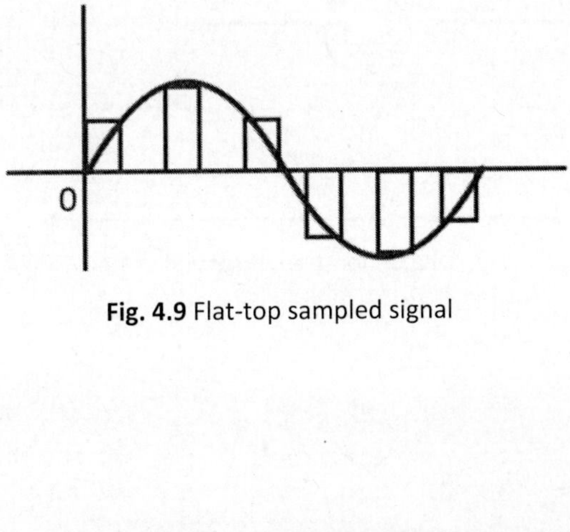

Fig. 4.9 Flat-top sampled signal

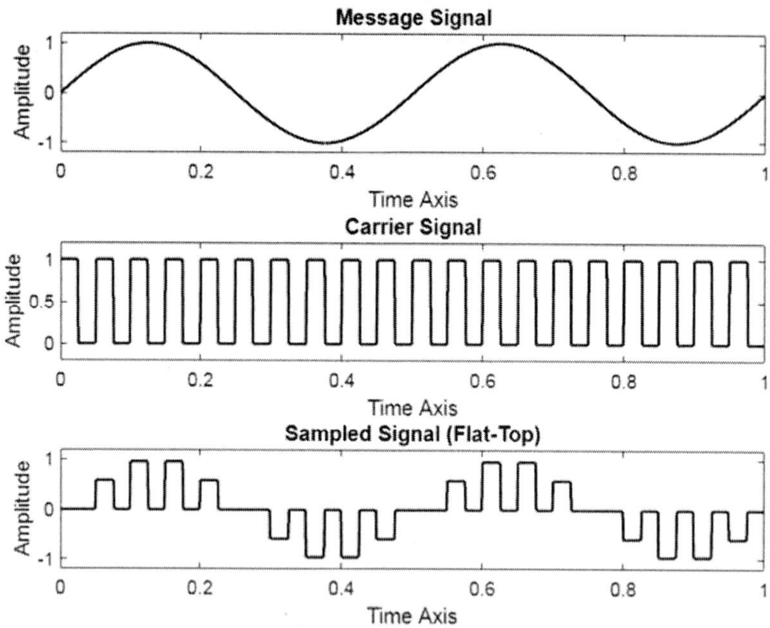

Fig. 4.10 Flat-top sampling

3. Natural Sampling

Natural sampling is also called practical sampling. In this sampling technique, the sampling signal is a pulse train. In natural sampling method, the top of each pulse in the sampled signal retains the shape of the input signal x(t) during pulse interval.

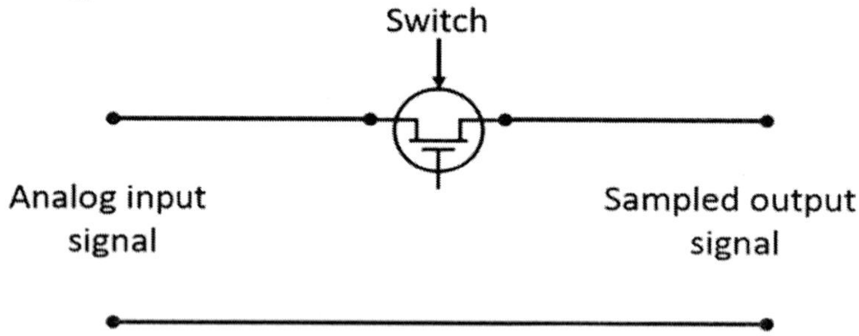

Fig. 4.11 Generation of natural sampled signal

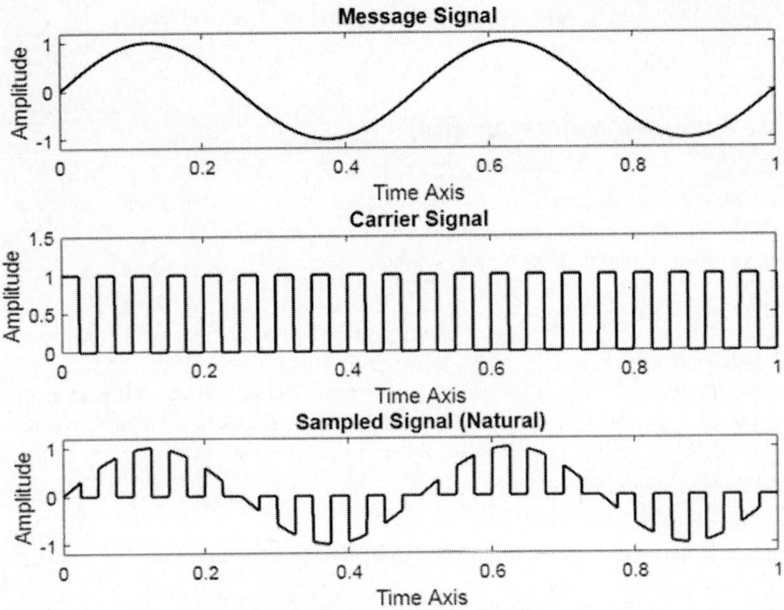

Fig. 4.12 Natural signal sampling

4.5.1 Aliasing

If signal is sampled at a rate lesser than the Nyquist Rate, the side band overlap, producing an interference-effect. This is called the Aliasing Effect as shown in Fig. (4.13). If aliasing takes place, it is not possible to recover the original analog signal.

The corrective measures taken to reduce the effect of Aliasing are –

- In the transmitter section of PCM, a low pass anti-aliasing filter can be employed before the sampler to eliminate the high frequency components, which are unwanted.

- After filtering, the signal should be sampled at a rate higher than the Nyquist rate. Having the sampling rate higher than Nyquist rate, also helps in the easier design of the reconstruction filter at the receiver.

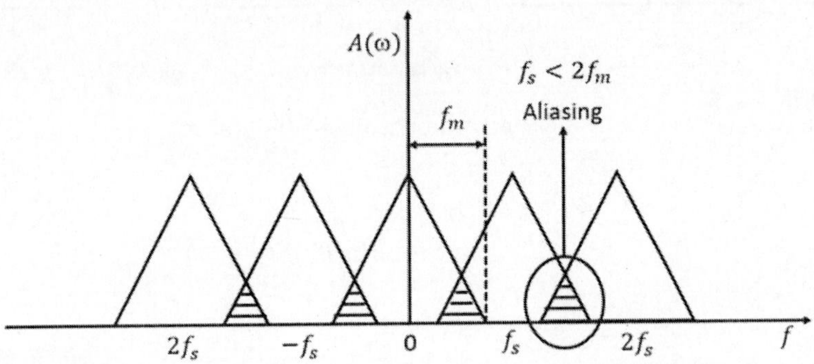

Fig. 4.13 Overlapping of spectrum due to aliasing

4.5.2 Inter Symbol Interference (ISI)

Inter-symbol interference (ISI) occurs when one symbol interferes with the other subsequent symbols as shown in Fig. (4.14). This is a form of distortion and an unwanted phenomenon. For any one symbol, all the remaining symbols have an effect similar to noise on that symbol.

Inter-symbol interference can also be caused due to multipath propagation. We know that wireless signal can take several different routes before it reaches a receiver. This happens when wireless signals refract through objects, bounce off surfaces, and are affected by air conditions. These pathways differ in length before they reach the recipient, resulting in many copies of the signal arriving at various times. Correct symbol detection will be hampered by the delay in symbol transmission. When the various pathways are received for extra interference, the signal's amplitude and/or phase may be altered.

Inter-symbol interference can be reduced with the help of adaptive equalization techniques and error correcting codes.

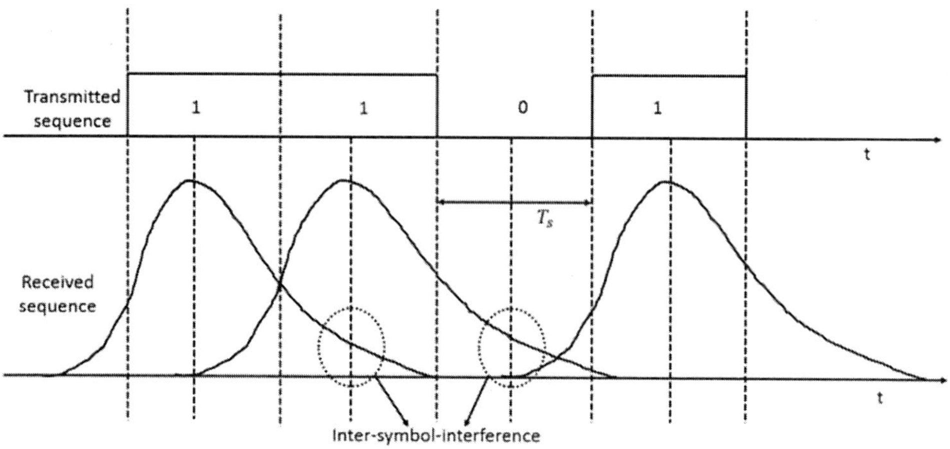

Fig. 4.14 Inter-symbol-interference

References and Suggested Reading

1. B. P. Lathi, and Zhi Ding, "Modern Digital and Analog Communication Systems", Oxford University Press, 2010.
2. Simon Haykin, "Communication Systems", Wiley, 2007.
3. Herbert Taub, Donald L. Schilling, and Gautam Saha, "Principal of Communication Systems", McGraw Hill Education, 2017.
4. P.F. Panter, "Modulation, Noise and Spectral Analysis", McGraw-Hill, New York, 1965.
5. https://nptel.ac.in/courses/117102059
6. https://ocw.mit.edu/courses/16-36-communication-systems-engineering-spring-2009/pages/lecture-notes/
7. George Clifford Hartley, "Techniques of Pulse-code Modulation in Communication Networks", 1967.

5 Passband Digital Transmission

UNIT SPECIFICS

Through this unit we have discussed the following aspects:

- *Pass band digital modulation Schemes,*
- *Phase shift keying,*
- *Frequency shift keying,*
- *Quadrature amplitude*
- *Continuous phase modulation,*
- *Minimum shift keying,*
- *Transmitter, receiver spectrum and signal space representation of passband modulation schemes.,*

The practical applications of the topics are discussed for generating further curiosity and creativity as well as improving problem solving capacity.

Besides giving a large number of multiple-choice questions as well as questions of short and long answer types marked in two categories following lower and higher order of Bloom's taxonomy, assignments through a number of numerical problems, a list of references and suggested readings are given in the unit so that one can go through them for practice. It is important to note that for getting more information on various topics of interest some QR codes have been provided in different sections which can be scanned for relevant supportive knowledge.

After the related practical, based on the content, there is a "Know More" section. This section has been carefully designed so that the supplementary information provided in this part becomes beneficial for the users of the book. This sectionj mainly highlights the initial activity, examples of some interesting facts, analogy, history of the development of the subject focusing the salient observations and finding, timelines starting from the development of the concerned topics up to the recent time, applications of the subject matter for our day-to-day real life or/and industrial applications on variety of aspects, case study related to environmental, sustainability, social and ethical issues whichever applicable, and finally inquisitiveness and curiosity topics of the unit.

RATIONALE

This unit help students to get an idea of different modulation schemes for passband transmission. We have started with basics and pre-requisites required for the topic. First, basic modulation schemes such as phase shift keying, frequency shift keying, quadrature amplitude modulation, is thoroughly explained before moving on to the more complex topics. Then we have covered the continuous phase modulation, with its transmitter, receiver, spectrum and signal space representation. Similarly, the other digital modulation schemes are discussed to provide an idea about the passband transmission. The chapter is concluded with some MCQs and numerical questions.

Digital Modulation is the most important step in transmitting an information bearing signal over a passband transmission. It accomplishes the task of shifting the range of frequencies of the signal to some another frequency range suitable for transmission. We will learn more about the modulation process in this chapter.

PRE-REQUISITES

Mathematics: Calculus (Class XII)

Physics: Basic of Communication (Class XII)

UNIT OUTCOMES

List of outcomes of this unit is as follows:

U1-O1: *Describe different passband digital modulation schemes,*

U1-O2: *Explain PSK, FSK, and QAM modulations,*

U1-O3: *Describe the continuous phase modulation, minimum shift keying,*

U1-O4: *Study of transmitter and receiver of different modulation schemes,*

U1-O5: *Spectrum and signal space representation for passband modulation schemes.*

This table below need to be filled after discussion as marked with red.

Unit-1 Outcomes	*EXPECTED MAPPING WITH COURSE OUTCOMES* (1- Weak Correlation; 2- Medium correlation; 3- Strong Correlation)					
	CO-1	*CO-2*	*CO-3*	*CO-4*	*CO-5*	*CO-6*
U1-O1	3	3	3	-	3	1
U1-O2	1	1	2	2	1	-
U1-O3	2	1	3	1	2	1
U1-O4	-	-	3	1	2	2
U1-O5	3	3	3	-	3	1

5.1 PHASE SHIFT KEYING

5.1.1 BPSK Modulation

The carrier's phase is shifted according to the input data stream, but its frequency and amplitude remain constant.

Binary PSK (BPSK): two phases (0° 180°) corresponds to two binary digits.

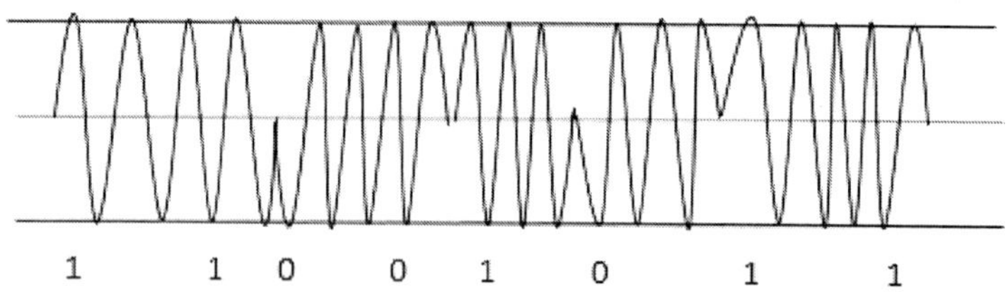

Fig. 5.1: BPSK waveform

Symbols 1 and 0 in binary PSK (BPSK) are represented by $S_1(t)$ $S_2(t)$ and are defined as

$$S_1(t) = \sqrt{\frac{2E_b}{T_b}}\cos[2\pi f_c t + 0^o] = \sqrt{\frac{2E_b}{T_b}}\cos[2\pi f_c t] \quad 0 \le t \le T_b \tag{5.1}$$

$$S_2(t) = \sqrt{\frac{2E_b}{T_b}}\cos[2\pi f_c t + 180^o] = -\sqrt{\frac{2E_b}{T_b}}\cos[2\pi f_c t] \quad 0 \le t \le T_b \tag{5.2}$$

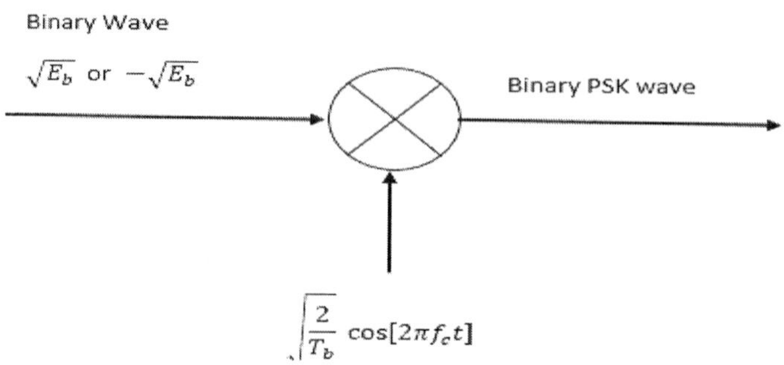

Fig.5.2 BPSK Transmitter

There is single basis function $\phi_1(t) = \sqrt{\frac{2}{T_b}}\cos[2\pi f_c t] \quad 0 \le t \le T_b$,

$$S_1(t) = \sqrt{E_b}\,\phi_1(t) \qquad 0 \le t \le T_b$$

$$S_2(t) = -\sqrt{E_b}\,\phi_1(t) \qquad 0 \le t \le T_b$$

The BPSK system has a one-dimensional signal space and two message points with the coordinate points shown below.

$$s_{11} = \int_0^{T_B} S_1(t)\phi_1(t)\,dt = \int_0^{T_B} \sqrt{\frac{2E_b}{T_b}}\cos[2\pi f_c t] \;*\; \sqrt{\frac{2}{T_b}}\cos[2\pi f_c t] \;=\; \sqrt{E_b} \qquad (5.3)$$

$$s_{21} = \int_0^{T_B} S_2(t)\phi_1(t)\,dt = \int_0^{T_B} -\sqrt{\frac{2E_b}{T_b}}\cos[2\pi f_c t] \;*\; \sqrt{\frac{2}{T_b}}\cos[2\pi f_c t] \;=\; -\sqrt{E_b} \qquad (5.4)$$

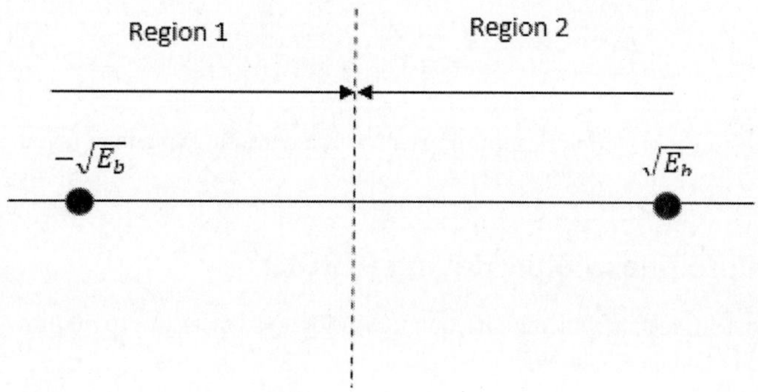

Fig. 5.3 Space signal diagram

The message $S_1(t)$ is located at $s_{11} = \sqrt{E_b}$ $S_2(t)$ is at $s_{21} = -\sqrt{E_b}$.

The signal space has two regions 1 and 2 corresponding to message 1 and 0.

The distance between message points is $2\sqrt{E_b}$

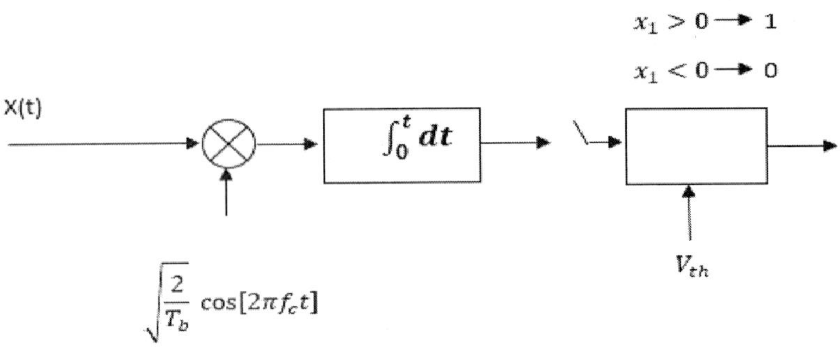

Fig.5.4 BPSK Receiver

$$P_e(0) = \frac{1}{2} \, erfc\left(\frac{E_b}{N_0}\right)$$

Similarly, if an error occurs while transmitting symbol 1, a decision is made in favour of symbol 0.

5.1.2 Quadrature phase Shift Keying (QPSK)

The important goal in design of a digital communication system is to provide low probability of error and efficient utilization of channel bandwidth.

In QPSK, the phase of the carrier takes on one of four equally spaced values, such as π/4, 3π/4 5π/4 7π/4 as given below

$$S_i(t) = \sqrt{\frac{2E}{T}} \cos\left[2\pi f_c t + (2i-1)\frac{\pi}{4}\right] \qquad 0 \le t \le T$$

where i =1,2,3,4, E is the transmitted signal energy per symbol, T is symbol duration, and the carrier frequency f_c equals n_c/T for some fixed integer n_c.

$$S_i(t) = \sqrt{\frac{2E}{T}} \cos\left[(2i-1)\frac{\pi}{4}\right] cos2\pi f_c t - \sqrt{\frac{2E}{T}} \sin\left[(2i-1)\frac{\pi}{4}\right] sin2\pi f_c t \qquad (5.5)$$

There are two orthonormal basis functions contained in $S_i(t)$

$$\phi_1(t) = \sqrt{\frac{2}{T}} \cos[2\pi f_c t] \qquad 0 \le t \le T$$

$$\phi_2(t) = \sqrt{\frac{2}{T}} \sin[2\pi f_c t] \qquad 0 \le t \le T$$

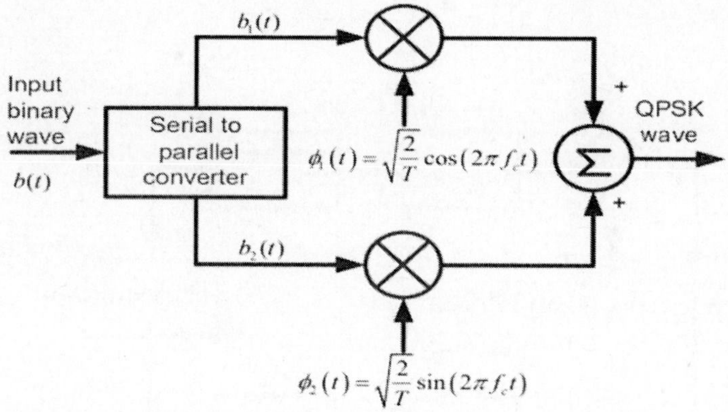

Fig. 5.5 QPSK Transmitter

Constellation Diagram

When we use two carriers, one in quadrature of the other, a constellation diagram can help us define the amplitude and phase of a signal. The in-phase carrier is represented by the X-axis, and the quadrature phase carrier is represented by the Y-axis.

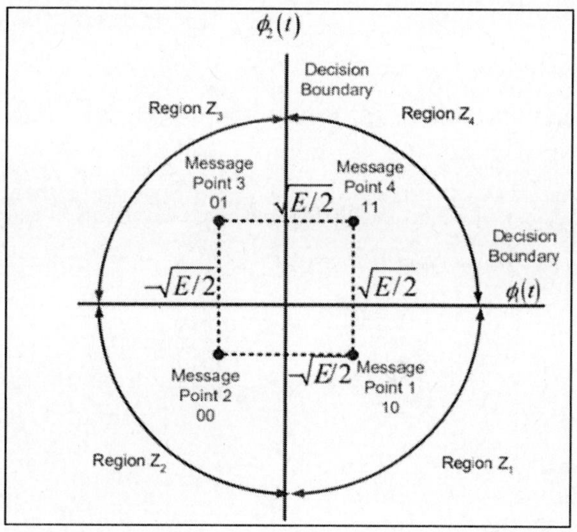

Fig. 5.6 QPSK constellation diagram

There are four message points,

$$S_i = \sqrt{E}\,\cos\left[(2i-1)\frac{\pi}{4}\right]$$

$$S_i = -\sqrt{E}\ \sin\left[(2i-1)\frac{\pi}{4}\right]$$

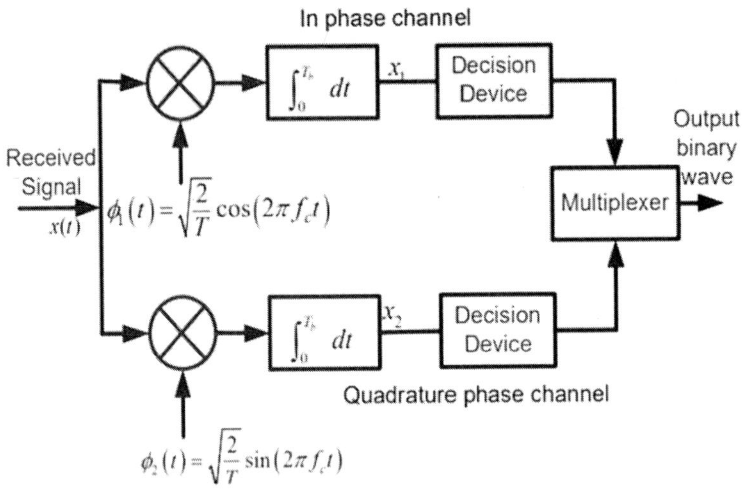

Fig. 5.7 QPSK receiver

The received signal is defined by

$$x(t) = S_i(t) + w(t) \qquad 0 \le t \le T \quad i = 1, 2, 3, 4$$

The observation vector x of a coherent QPSK receiver has 2 elements x1 and x2 and are defined by probability of symbol error of QPSK,

$$P_e = \frac{1}{2}\, erfc\left(\sqrt{\frac{E}{2N_0}}\right) \qquad (5.6)$$

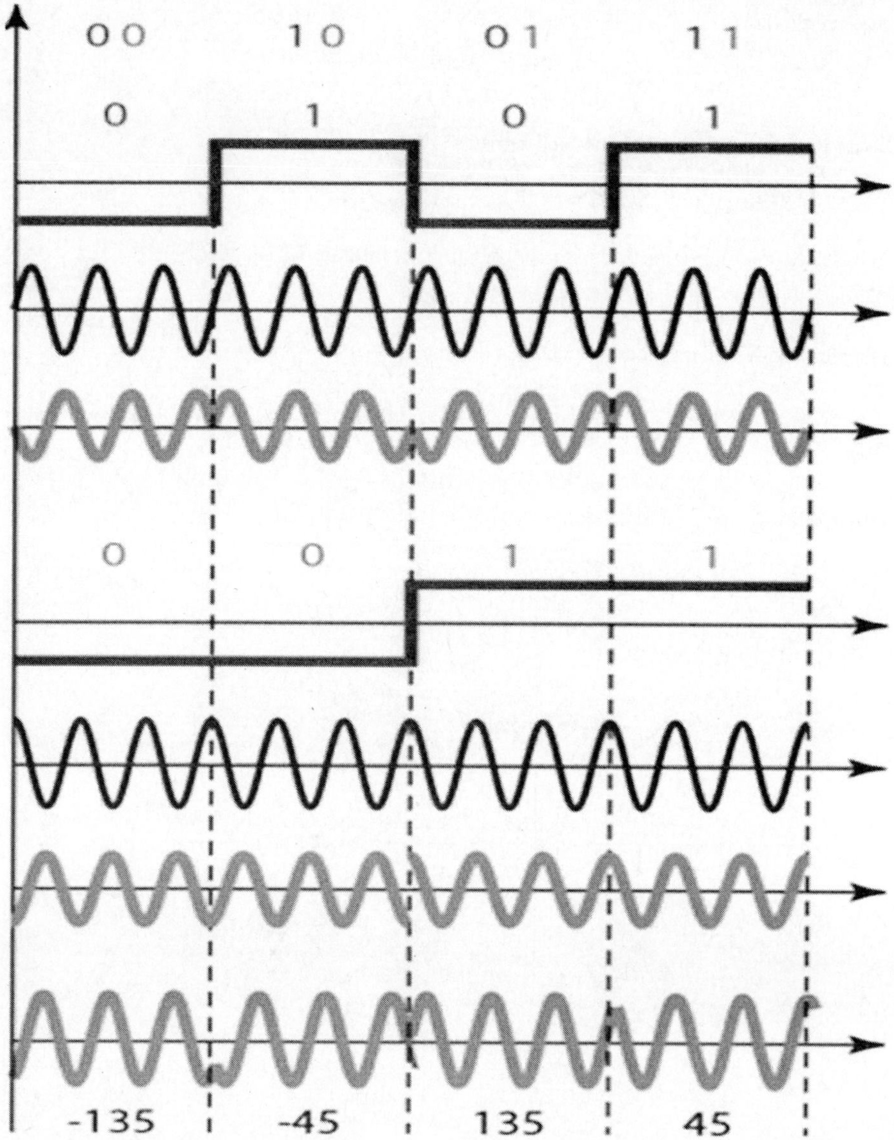

Fig. 5.8 QPSK waveform

5.1.3 M-ary PSK

High data rates are possible with multi-level modulation techniques within fixed bandwidth constraints. A useful set of signals for M-ary PSK is

$$\emptyset_i = A\cos(\omega_c t + \theta_i) \qquad 0 < t \leq T_s$$

where the M phase angles are

$$\theta_i = 0, \frac{2\pi}{M}, \dots, \frac{2(M-1)\pi}{M}.$$

The PSD for M-ary PSK for equiprobable ones and zeros is

$$S_\phi(\omega) = A^2 T_s Sa^2\left[(\omega - \omega_c)\frac{T_s}{2}\right]$$

The symbols in this case are of duration T_s , so the information (or bit) rate T_b satisfies

$$T_s = T_b \log_2 M$$

M-ary PSK's potential bandwidth efficiency can be demonstrated

$$\frac{f_b}{B} = \log_2 M \; bps/Hz$$

For the case of M = 8, a phase diagram and signal constellation diagram are shown below

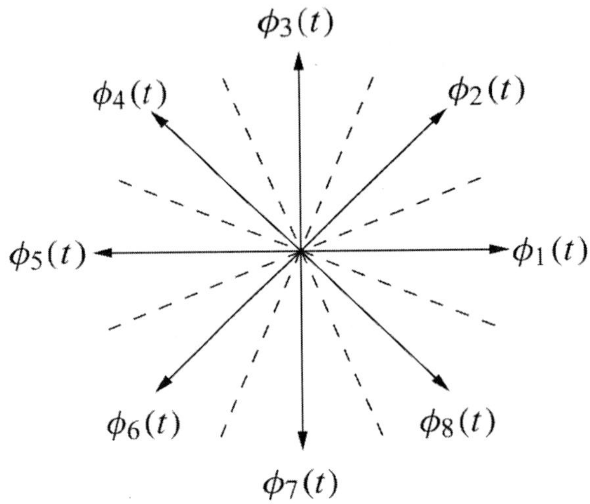

Fig. 5.9 M-ary PSK phase diagram and signal constellation

All signals have the same energy E_s over the interval (0, T_s), and each signal is correctly demodulated at the receiver if the phase is within $\pm\pi/M$ of the correct phase θ_i . No information is contained in the energy of the signal. A probability of error calculation involves analysing the received phase at the receiver (in the presence of noise) and comparing it to the actual phases. An exact solution is difficult to compute, but for $P_\epsilon < 10^{-3}$ approximate probability of making a symbol error is

$$P_\epsilon \approx 2erfc\sqrt{\frac{2E_s}{\eta}sin^2\frac{\pi}{M}} \qquad M > 2$$

When using a Gray code, the corresponding bit error is approximately

$$P_{be} \approx P_\epsilon / \log_2 M$$

Stemler provides a table of M-ary PSK SNR requirements for fixed error rates. The results show that QPSK (M = 4) has a clear advantage over coherent PSK (M = 2) in terms of bandwidth efficiency, with only a 0.3dB increase in SNR. M = 8 is frequently used for higher-rate transmissions in bandlimited channels. M > 8 values are rarely used due to the high-power requirements.

M-ary PSK signaling necessitates more sophisticated equipment than BPSK signaling. Carrier recovery is also more difficult. The requirement for the carrier to be recovered can be reduced by comparing the phases of two successive symbols. This results in M-ary differential PSK, which is similar to DPSK (which is differential PSK for M = 2).

For large SNR the probability of error is

$$P_\epsilon \approx 2erfc\sqrt{\frac{2E_s}{\eta}sin^2\frac{\pi}{\sqrt{2M}}}$$

Thus, differential detection increases the power requirements by the factor

$$\Gamma = \frac{sin^2\pi/M}{sin^2(\pi/\sqrt{2M})}$$

For $m = 4$, the increase in required power is about 2.5dB, which may be justified by the saving in equipment complexity

5.2 FREQUENCY SHIFT KEYING

A technique for transferring digital signals utilizing discrete signals is called frequency shift keying (FSK). We have two different frequency signals according to binary symbols since the carrier's frequency is changed while the carrier's phase is unaffected

Let there be a frequency shift by ω

If $\mathrm{b(t)} = 1;\qquad \mathrm{s(t)} = \sqrt{2}\mathrm{p}\cos(2\pi f_0 + \omega)\mathrm{t}$

$\mathrm{b(t)} = 0;\qquad \mathrm{s(t)} = \sqrt{2}\mathrm{p}\cos(2\pi f_0 - \omega)\mathrm{t}$

Thus, when symbol 1 is to be transmitted the carrier frequency will be $f_0 + \frac{\omega}{2\pi}$

and if symbol 0 is transmitted the carrier frequency will be $f_0 - \frac{\omega}{2\pi}$

5.2.1 FSK TRANSMITTER

The input binary sequence is coupled to two oscillators with a clock in the FSK modulator block diagram. To prevent phase discontinuities of the output waveform during message transmission, two oscillators that produce higher and lower frequency signals are coupled to a switch along with an internal clock. Both oscillators have clocks applied internally, which allows the transmitter to select the frequencies in accordance with the binary input sequence.

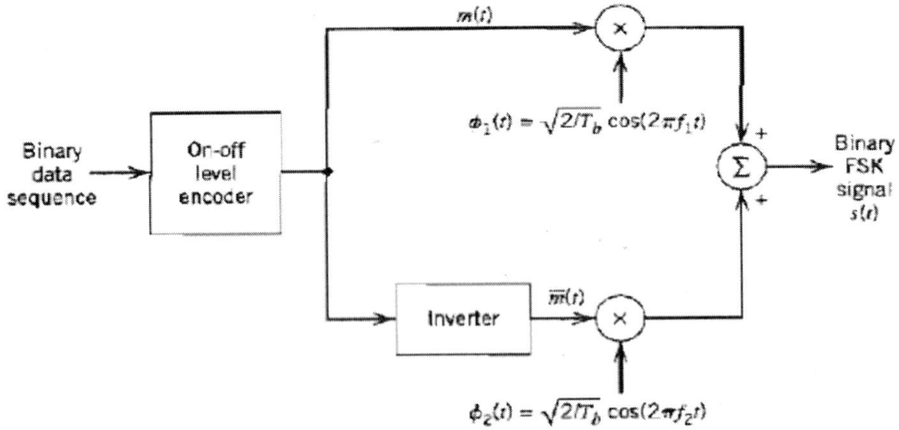

Fig 5.10 FSK Transmitter

5.2.2 FSK RECEIVER

A FSK wave can be demodulated in several ways. Asynchronous and synchronous detectors are the primary means of FSK detection. In contrast to the non-coherent asynchronous detector, the synchronous detector is coherent one.

Asynchronous detector: The block diagram shows an asynchronous FSK detector with two envelope detectors, two band pass filters (BPF), and a decision circuit. The output from these two BPFs, which are pass filters set to space and marking frequency, resembles the ASK signal that is provided to the envelop detector. The decision circuit selects the most likely output from among all the envelope detectors. Additionally, it gave the waveform a rectangular shape.

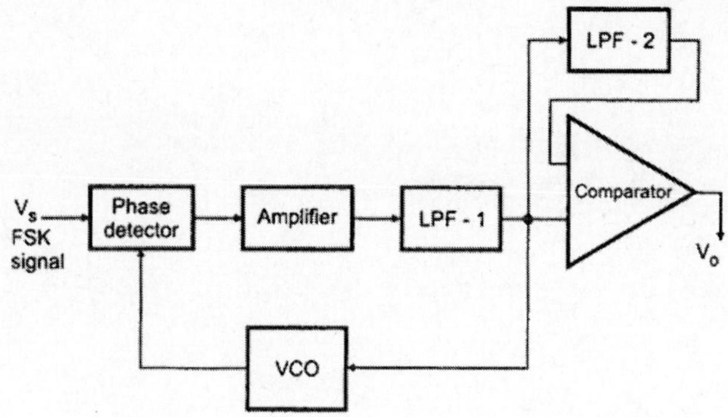

Fig 5.11 Frequency shift keying demodulator

Synchronous detector:

A decision circuit, two band pass filters, two mixers with local oscillator circuits, and two FSK detectors are shown in the synchronous FSK detector's block diagram. Two mixing circuits with oscillators receive the FSK signal input. The decision circuit determines which output is more likely and chooses it from either one of the detectors; the two signals have a minimum frequency separation. These two are coupled to two band passes filters; together, they serve as a demodulator. Synchronous demodulators are a little more complicated than asynchronous demodulators in that the bandwidth of each of them depends on their respective bit rates.

5.2.3 SPECTRUM OF FSK

FSK switches between two carrier frequencies. Analysing two coexisting frequencies is easier. The FSK spectrum is created by merging two ASK spectra that are centred on the bandwidth required for FSK transmission, which is equivalent to the signal's band rate plus the difference in the two carriers' frequencies.

$$BW = (f_{c_1} - f_{c_0}) + N\ band$$

Even though there are only two carrier frequencies, the modulation phase produces a composite signal made up of numerous simple signals, each with a variety of frequencies.

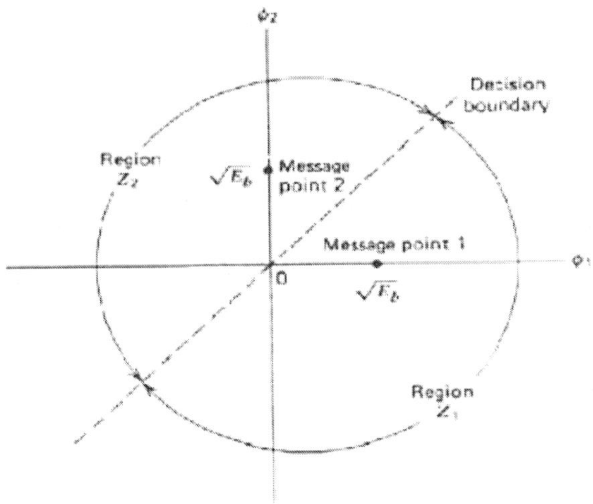

Fig 5.12 Signal space diagram of coherent binary FSK system

5.2.4 M-ary Frequency Shift Keying

A power-efficient modulation technique called M-ary frequency shift keying (FSK) exhibits an improvement in performance as the number of frequencies used (M) rises, but at the cost of greater complexity and lower bandwidth efficiency. This strategy has been proved to be beneficial in low rate, low power applications. However, there is a divergence between the performance of theoretical M-ary FSK systems and that of systems using M-ary FSK modulation schemes in the real world.

Modulation

It is simple to convert the coherent modulator of binary FSK to coherent M-ary FSK shown in figure below. Here, M signals with the intended frequencies and coherent phase are produced using the frequency synthesizer. The multiplexer selects a frequency based on the $n = \log_2 M\ bits$.

A general type of detector for M-ary equal-energy & equiprobable signals with known phases is the coherent M-ary FSK demodulator. The demodulator is made up of a bank of M matched filters, as indicated in the figures below. The receiver makes decisions based on the maximum output of the correlators or matched filters at sample times t = kT. It is important to note that the coherent M-ary

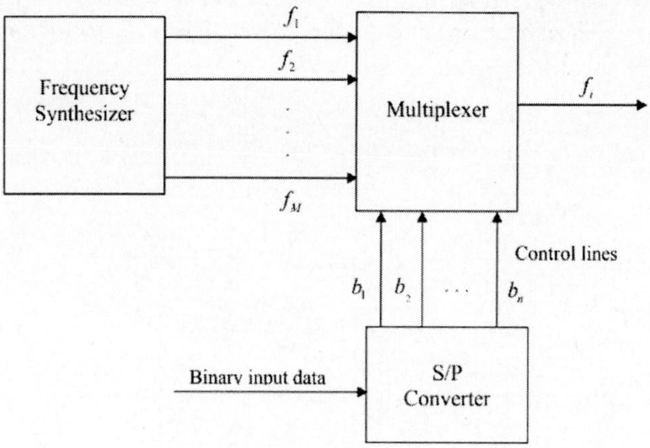

Coherent M-ary FSK modulator.

Fig.5.13 Coherent M-ary FSK modulator

FSK receivers shown in the figures below essentially require that the M-ary FSK signals be equiprobable and equal in energy not necessarily orthogonal.

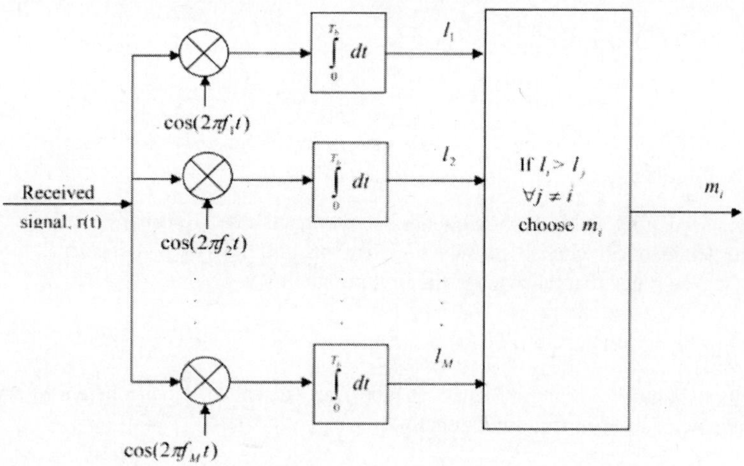

Fig.5.14 Coherent M-ary FSK demodulator correlator implementation

For a symmetrical signal set with equal energy and equiprobable, the accurate formula for the symbol error probability is given as

$$P_s = 1 - \int_{-\infty}^{\infty} \frac{1}{\sqrt{2\pi}} exp\left\{-\frac{\left(x - \sqrt{2E_s/N_0}\right)^2}{2}\right\}[1 - Q(x)]^{M-1}dx$$

This expression cannot be evaluated analytically since it does not require an orthogonal signal set. All separations between any two signals are equal if the signal set is equal-energy and orthogonal. The upper bound determined by the equation above for distance $d = \sqrt{2E_s}$ is

$$P_s \leq (M-1)Q\left(\sqrt{\frac{E_s}{N_0}}\right)$$

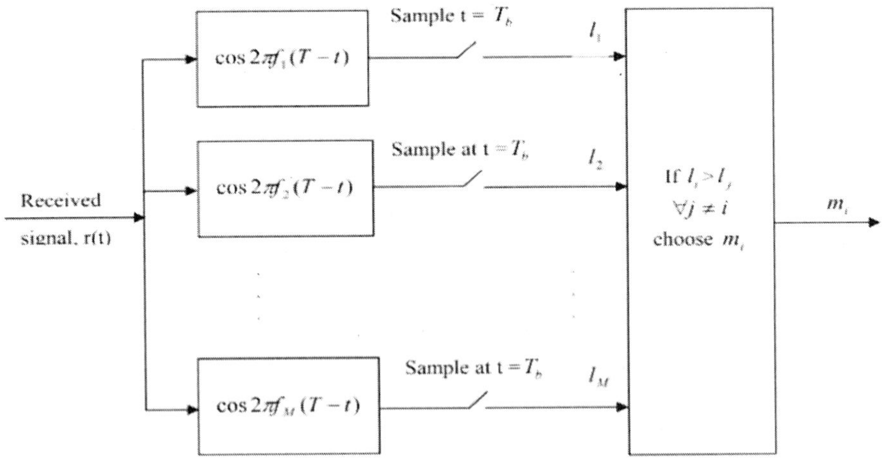

Fig.5.15 Coherent M-ary FSK demodulator matched filter

where $E_s = E_b \log_2 M$, and $Q(z)$ is the Q-function. As E_s/N_0 is increased for fixed M, this constraint gets increasingly tight. It turns into a reliable approximation for $P_s \leq 10^{-3}$

All symbol errors are equiprobable for orthogonal M-ary signals that are equally likely. In other words, the demodulator has an identical chance of selecting any one of the (M −1) incorrect orthogonal signals. Consequently, it can be stated that the average bit error probability is given by

$P_b = \frac{2^{n-1}}{2^n - 1} P_s = \frac{2^{n-1}}{M-1} P_s$, where $n = \log_2 M$ $bits$.

The noncoherent modulator for binary FSK can also be easily extended to noncoherent M-ary FSK by simply increasing the number of independent oscillators to M.

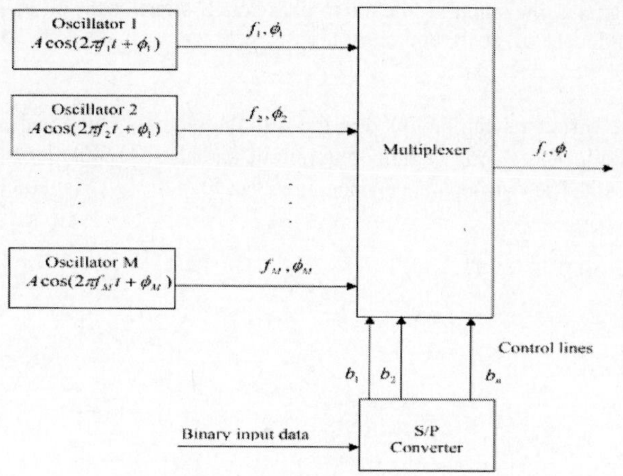

Fig.5.16 Noncoherent M-ary FSK modulator

The demodulator can be implemented as a matched filter-squarer, matched filter-envelope detector, or correlator-squarer.

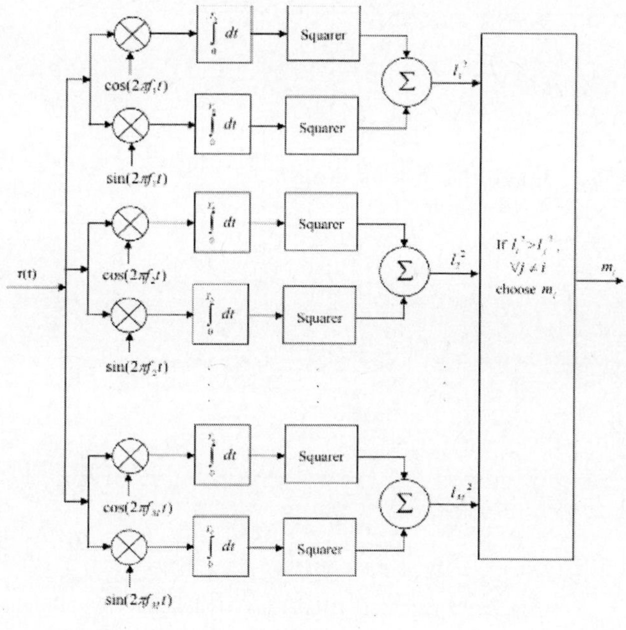

Fig.5.17 Noncoherent M-ary FSK demodulator correlator implementation

5.3 QUADRATURE AMPLITUDE MODULATION (QAM)

Two carriers (sine and cosine), which are phase-shifted by 90 degrees are modulated and combined to form this signal known as QAM. They are in quadrature as a result of their 90° phase difference, hence the name. The quadrature or "Q" signal is one signal, while the in-phase or "I" signal is the other. QAM, is a type of modulation that can offer high levels of spectrum consumption efficiency, makes use of both amplitude and phase components.

Phase and amplitude variations can be found in the resulting signal, which is the combination of both I and Q carriers. It can also be seen as a combination of amplitude and phase modulation since both amplitude and phase fluctuations are present. The two separate carrier signals used in the QAM method are shown in the waveforms below Fig. (5.18).

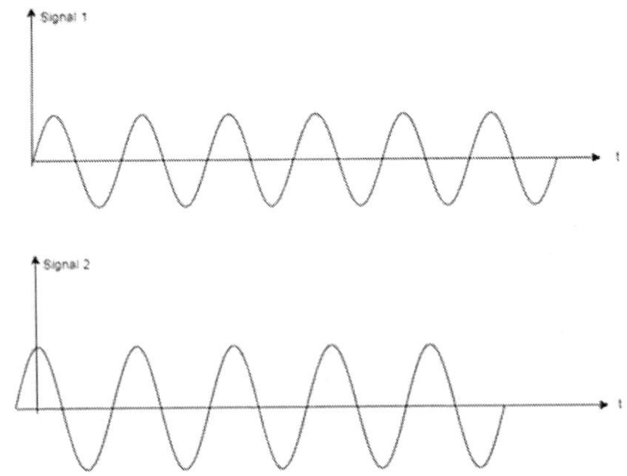

Fig 5.18 Sinusoidal carrier signals

The output waveform of QAM is shown below in Fig 5.19

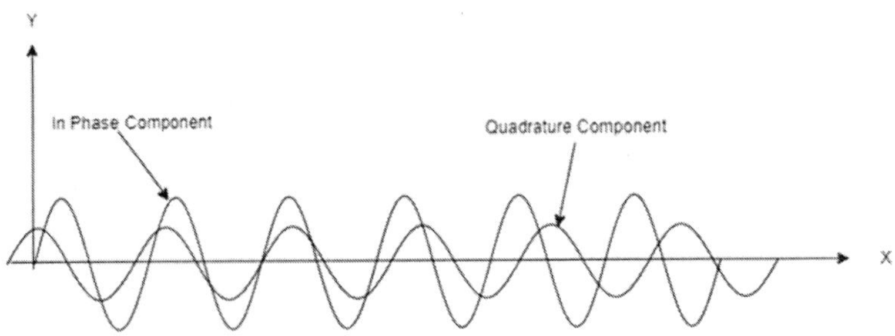

Fig 5.19 Quadrature Amplitude Modulation Output Signal

5.3.1 QAM Transmitter

The aforementioned segment, which consists of a product modulator 1 and a local oscillator, is referred to as the In-phase channel in a QAM transmitter, whereas a product modulator 2 and local oscillator are referred to as the quadrature channel. The output is a QAM as a result of adding the output signals from the in-phase and quadrature channels.

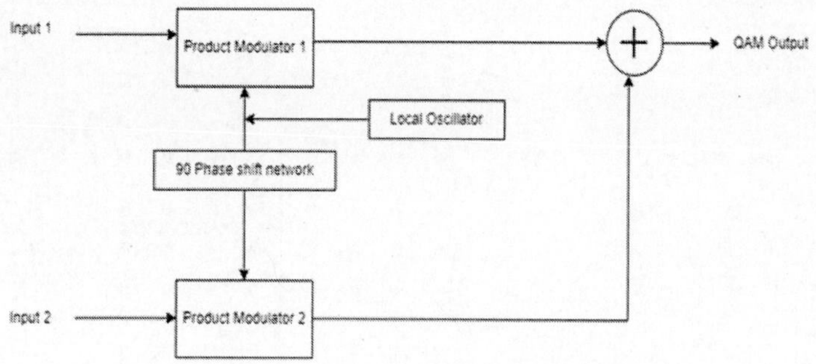

Fig **5.20** Quadrature Amplitude Modulation transmitter

5.3.2 QAM Receiver

At the receiver level, the QAM signal is forwarded from the upper channel of receiver and lower channel, and the resultant signals of product modulators are forwarded from LPF1 and LPF2. The cut off frequencies of input 1 and input 2 signals are fixed, so the recovered original signals are the filtered outputs from these filters.

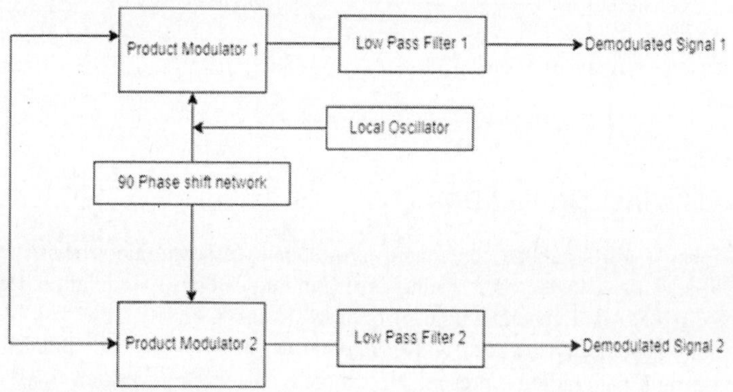

Fig. 5.21 Quadrature Amplitude Modulation receiver

5.3.3 Constellation Diagrams

The quality and distortion of a digital transmission are represented graphically using constellation diagrams. Higher data rates can be achieved with Quadrature Amplitude Modulation (QAM), but at the expense of the noise margin. QAM uses a variety of phases known as states, including 16, 32, 64, and 256 and each state is defined by a specific amplitude and phase.

16 QAM and 64 QAM are shown below in Fig (5.22)

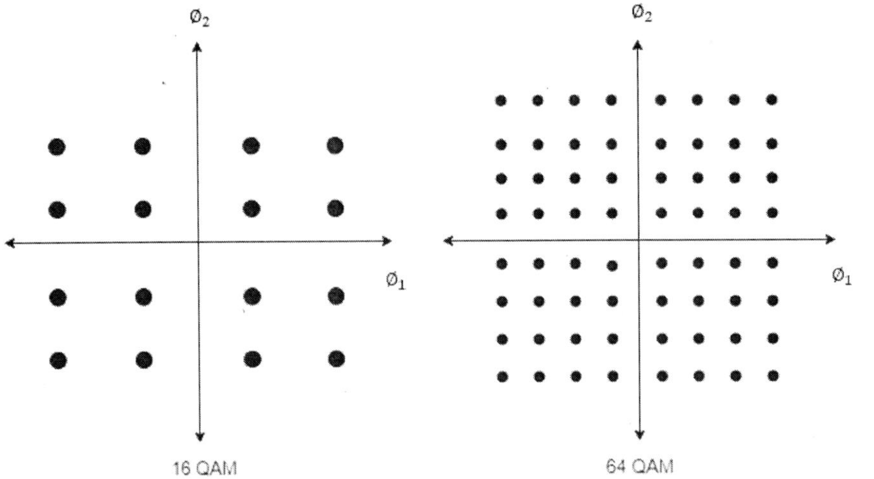

Fig. 5.22 16 QAM and 64 QAM

Energy Of Constellation and minimum distance -

Average energy of a constellation = $\frac{Total\ energy\ of\ all\ symbols}{Total\ number\ of\ symbols}$, where energy of a symbol = $radius^2$,

And radius = distance of symbol from origin.

Minimum distance between two symbols is denoted hai d_{min}.

5.4 Continuous Phase Modulation

Minimum shift-keying (MSK) can be considered as a subpart of continuous phase frequency shift-keying (CPFSK) and CPFSK can be considered as a subpart of Continuous Phase Modulation. Hence continuous phase modulation scheme (CPM) is a generalization of both MSK and CPFSK. CPM is a non-linear and constant amplitude modulation scheme. It has been widely applied in such fields of communication, where nonlinear distortions as well as multipath fading is present. Here there is a need for a constant signal envelope and efficient bandwidth necessary.

In CPM the carrier phase varies in a continuous manner while the envelope of the output signal is constant. It is a very spectrally efficient method because the lack of phase discontinuities ensures the minimization of the high-frequency spectral components. CPM is considered both bandwidth as well as power efficient modulation technique.

The multi-h phase coding method of CPM, where h denotes the modulation index, is one that can result in significant coding gains. The transmission of the same symbol over two consecutive symbol intervals causes various phase changes. Hence CPM is a generalization of the CPFSK schemes. The expression $iT \leq t(i+1)T$ can be used to represent the signal during the ith interval mathematically:

$$x(t,d) = \sqrt{\frac{E_s}{T}} cos(w_c t + d_i w_i (t - iT) + \varphi_i) \quad (5.7)$$

where, E_s is the energy of the symbol, T is the symbol duration, w_c is the carrier frequency in radians/second, and d and ω are the sequences that represent the M-ary information sequence.

The phase associated with the current data symbol and the phase accumulation due to the previous data symbol are given by the data phase terms $d_i w_i(t - iT)$ and φ respectively,

$$d_i w_i(t - iT) \ = d_i \int_{iT}^{T} [(\pi h_i)/T]\, g(t1 - iT) dt_1$$

$$\varphi_i \ = \ \sum_{k=0}^{i-1} \pi d_k h_k \qquad (5.8)$$

where w_i is the angular frequency associated with h which is the modulation index used during the ith baud. In a round-robin approach, the various values of h can be used between symbol intervals.

A type of frequency modulation known as minimum shift keying, or MSK, is based on the continuous-phase frequency-shift keying which is a generalization of CPM. As a result MSK is also a type of continuous phase modulation. When compared to other modes of a similar kind, MSK has an advantage in terms of spectral efficiency. It also allows power amplifiers to operate in saturation, which helps increases their efficiency.

5.4.1 Transmitter

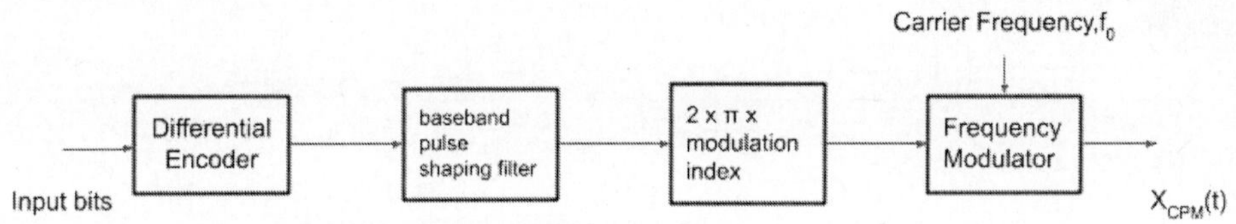

Fig. 5.23 Transmitter block diagram for continuous phase modulation

A general model of the transmitter block diagram for continuous phase modulation is shown in Fig. (5.23). Here, the input bits are fed into the differential encoder. The differential encoding is utilized to prevent the propagation of any error into the detector. G(f) denotes the baseband pulse shaping filter. It is selected in such a way that it generates narrow-band, elongated and partial-response pulses. The modulation index, h is chosen keeping in mind the trade-off between a narrow spectrum and a good detectability. The transmitted signal $X_{CPM}(t)$ is then passed to the channel.

5.4.2 Receiver

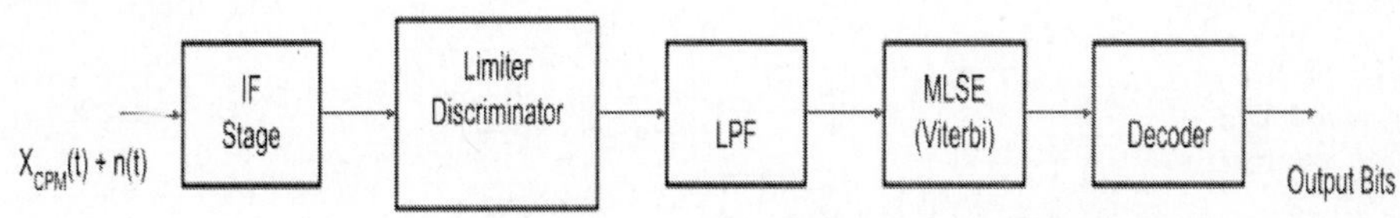

Fig. 5.24 Receiver block diagram for continuous phase modulation

On passing through the channel noise gets added to the CPM signal. At the receiver $X_{CPM}(t) + n(t)$ is detected as given in Fig. (5.24). It passes through an IF stage. IF filter compensates for the complex response of the high-order filters, typical of a tuner IF. The next stage is the Limiter-discriminator. In certain systems where phase coherence is very difficult to establish and/or maintain this detection of Continuous Phase Modulation (CPM) partial response signals plays an important role. The bandwidth and order of filter chosen for both the IF and LPF filters are dependent on a trade-off between detectability and crosstalk. The next stage comprises of a maximum likelihood sequence estimator or MLSE. Here we assume that the discriminator output comprises of white gaussian noise. MLSE when used significantly improves the detectability performance.

5.4.3 Spectrum

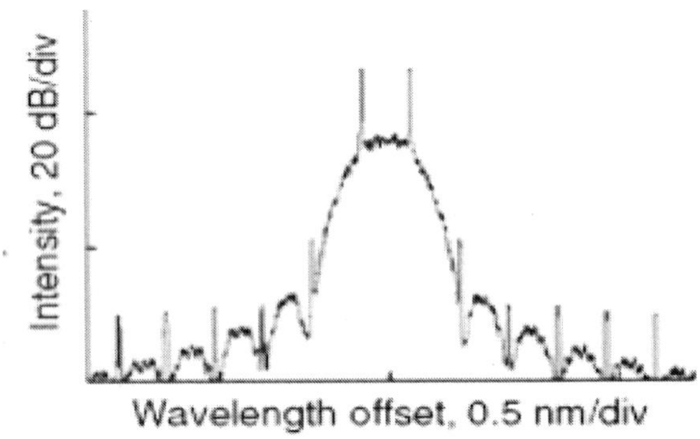

Fig. 5.25 continuous phase modulation spectrum

The spectra of the CPM signal resemble a sinc function. The power spectral density is compact in nature. It has a narrow main lobe and fast sidelobe roll-off. If we inspect the jphase transition diagram of PSK and FSK the phase transition is 180 degrees but in case of CPM the phase transition is smooth and minimum.

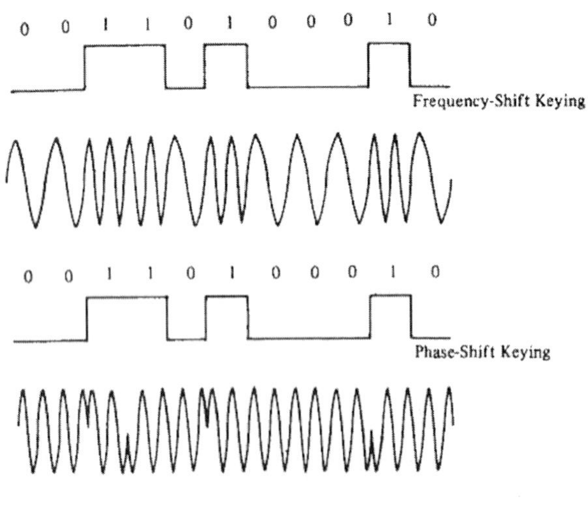

Fig. 5.26 PSK and FSK waveform diagram

5.5 MINIMUM SHIFT KEYING (MSK)

The phase information in the received signal is only partially utilised in the coherent detection of binary FSK signals to synchronise the receiver and transmitter. We now demonstrate that it is possible to considerably enhance the receiver's noise performance by properly utilising the phase when doing detection. However, this enhancement comes at the expense of a more sophisticated receiver. Consider a continuous-phase frequency-shift keying (CPFSK) signal, which is defined for the interval $0 \le t \le T_b$ as follows:

$$s(t) = \begin{cases} \sqrt{\frac{2E_b}{T_b}} \cos[\cos 2\pi f_1 t + \theta(0)] \; for \; symbol \, 1 \\ \sqrt{\frac{2E_b}{T_b}} \cos[\cos 2\pi f_2 t + \theta(0)] \; for \; symbol \, 0 \end{cases} \quad (5.9)$$

where E_b is the energy of transmitted signal per bit and T_b is the bit duration. The phase $\theta(0)$, represents the phase at time $t = 0$, sums up the history of the modulation process up to time $t = 0$. The frequencies f_1 and f_2 are transmitted with respect to binary symbols 1 and 0, respectively. The CPFSK signal $s(t)$ can also be expressed as a typical angle modulated signal as:

$$s(t) = \sqrt{\frac{2E_b}{T_b}} \cos[\cos 2\pi f_c t + \theta(t)] \quad (5.10)$$

where $\theta(t)$ represents the phase of s(t) at time t. When the phase $\theta(t)$ is a time dependent continuous function, we find that the modulated signal s(t) is also continuous with time, considering the inter-bit switching intervals. In each period of T_b seconds, the phase $\theta(t)$ of a CPFSK signal increases or drops linearly, as demonstrated by

$$\theta(t) = \theta(0) \pm \frac{\pi h}{T_b} t, \;\; 0 \le t \le T_b \quad (5.11)$$

where the plus sign is for transmitting symbol 1 and the minus sign for transmitting symbol 0. Putting (5.11) into (5.10), and then matching the angle from (5.9), we obtain the following pair of equations:

$$f_c + \frac{h}{2T_b} = f_1 \quad (5.12)$$

$$f_c - \frac{h}{2T_b} = f_2 \quad (5.13)$$

Evaluating for f_c and h, we obtain

$$f_c = \frac{1}{2}(f_1 + f_2) \quad (5.14)$$

and,

$$h = T_b(f_1 - f_2) \quad (5.15)$$

Therefore, the nominal carrier frequency f_c is the arithmetic mean of the transmitted frequencies f_1 and f_2. The deviation ratio h is defined as the difference between the frequencies f_1 and f_2, normalized with respect to the bit rate $\frac{1}{T_b}$.

With h = 1/2, we find from (7) that the frequency deviation (i.e., the difference between the two signalling frequencies f_1 and f_2) equals half the bit rate; hence the lowest frequency spacing that enables the two FSK signals representing symbol 1 and 0 to be coherently orthogonal is the frequency deviation h=1/2.

In other terms, symbols 1 and 0 do not interfere with one another in the process of detection. Therefore, for a deviation ratio of 0.5 the CPFSK signal is termed as minimum shift-keying (MSK).

5.5.1 Signal Space Diagram of MSK

We may rewrite the CPFSK signal S(t) in (5.10) in terms of its in-phase and quadrature components as:

$$s(t) = \sqrt{\frac{2E_b}{T_b}}\cos[\theta(t)]\ \cos[2\pi f_c t] - \sqrt{\frac{2E_b}{T_b}}\sin[\theta(t)]\ \sin[2\pi f_c t] \tag{5.16}$$

First, we consider the in-phase component $\sqrt{\frac{2E_b}{T_b}}\cos\theta(t)$, with the deviation ratio 0.5, from (5.11) we get:

$$s(t) = \sqrt{\frac{2E_b}{T_b}}\cos[\theta(t)]\ \cos[2\pi f_c t] - \sqrt{\frac{2E_b}{T_b}}\sin[\theta(t)]\ \sin[2\pi f_c t] \tag{5.17}$$

where the plus sign is for transmitting symbol 1 and the minus sign for transmitting symbol 0.
Same expression is true for θ(*t*) in the interval $-T_b \le t \le 0$, except for algebraic sign which may not be the same in both intervals. We discover that in the interval $-T_b \le t \le T_b$, the polarity of cosθ(*t*) depends only on θ(0), regardless of the sequence of 1s and 0s broadcasted
before or after *t* = 0. This is because the phase θ(0) is 0 or π depending on the
history of the modulation process. The half-cycle cosine pulse is thus the in-phase component for this interval:

$$\begin{aligned} s_I(t) &= \sqrt{\frac{2E_b}{T_b}}\cos[\theta(t)] \\ &= \sqrt{\frac{2E_b}{T_b}}\cos[\theta(0)]\ \cos\left(\frac{\pi}{2T_b}t\right) \\ &= \pm\sqrt{\frac{2E_b}{T_b}}\cos\left(\frac{\pi}{2T_b}t\right), \qquad -T_b \le t \le T_b \end{aligned} \tag{5.18}$$

Where the plus sign for $\theta(0) = 0$ and the minus sign for $\theta(0) = \pi$. Similar to this, we can demonstrate that, the *half-cycle sine pulse* constitutes the quadrature component of
s(*t*) in the interval $0 \le t \le 2T_b$:

$$\begin{aligned} s_Q(t) &= \sqrt{\frac{2E_b}{T_b}}\sin[\theta(t)] \\ &= \sqrt{\frac{2E_b}{T_b}}\sin[\theta(T_b)]\ \sin\left(\frac{\pi}{2T_b}t\right) \\ &= \pm\sqrt{\frac{2E_b}{T_b}}\sin\left(\frac{\pi}{2T_b}t\right), \qquad 0 \le t \le 2T_b \end{aligned} \tag{5.19}$$

where the plus sign for $\theta(T_b) = {}^{\pi}/_{2}$ and the minus sign for
$\theta(T_b) = -{}^{\pi}/_{2}$. Additionally, since each of the phase states θ(0) and θ(*T*b) may take one of the two potential values, any one of the following four outcomes is possible:

- $\theta(0) = 0$ and $\theta(T_b) = \pi/2$, which occur when sending symbol 1.

- $\theta(0) = \pi$ and $\theta(T_b) = \pi/2$, which occur when sending symbol 0.
  $\theta(0) = \pi$ and $\theta(T_b) = -\pi/2$ (or, equivalently, $3\pi/2$ modulo 2π), which occur when sending symbol 1.
- $\theta(0) = 0$ and $\theta(T_b) = -\pi/2$, which occur when sending symbol 0.

According to the values of the phase-state combination in this fourfold scenario, the MSK signal itself can take one of four different forms, corresponding to different θ(0) and θ(*T*b).

We can see from the expansion of (5.16), that the generation of MSK is characterized by two orthonormal basis functions $\phi_1(t)$ and $\phi_2(t)$, which are defined by the following pair of sinusoidally modulated quadrature carriers:

$$\phi_1(t) = \sqrt{\frac{2}{T_b}} \cos\left(\frac{\pi}{2T_b} t\right) \cos(2\pi f_c t), \ \ 0 \le t \le T_b \tag{5.20}$$

$$\phi_2(t) = \sqrt{\frac{2}{T_b}} \sin\left(\frac{\pi}{2T_b} t\right) \sin(2\pi f_c t), \ \ 0 \le t \le T_b \tag{5.21}$$

Correspondingly, we may write the MSK signal in the expanded form

$$s(t) = s_1\phi_1(t) + s_2\phi_2(t), \ \ 0 \le t \le T_b \tag{5.22}$$

where the phase states θ(0) and θ(*T*b), are linked to the coefficients s_1 and s_2, respectively. To obtain *s*1, the product *s*(*t*)ϕ1(*t*) is integrated with respect to time *t* between the limits –*T*b and *T*b, yielding

$$\begin{aligned} s_2 &= \int_0^{2T_b} s(t)\phi_2(t)dt \\ &= -\sqrt{E_b}\cos[\theta(T_b)], \qquad 0 \le t \le 2T_b \end{aligned} \tag{5.23}$$

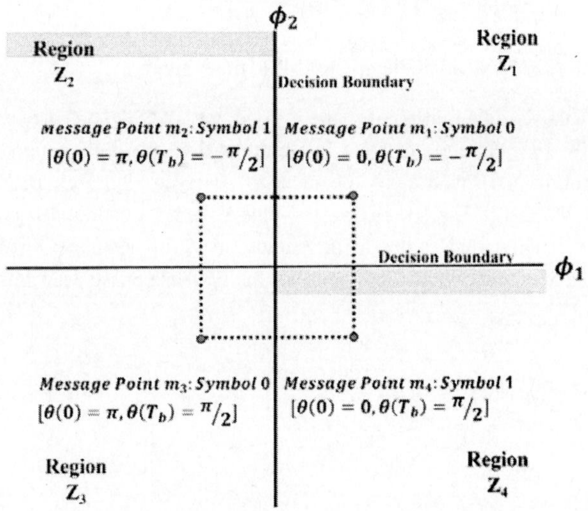

Fig. 5.27 Signal-space diagram for MSK system

Therefore, Fig. (5.27) depicts the signal space diagram, where the signal constellation for an MSK signal is two-dimensional (i.e., $N = 2$), with four potential signal points (i.e., $M = 4$).

5.5.2 MSK Transmitter and Receiver

Consider next the generation and demodulation of MSK. Fig. (5.28) shows the block diagram of a typical MSK transmitter. The benefit of using this technique to create MSK signals is that changes in the input data rate have no impact on the signal coherence and deviation ratio. First product modulator receives two input sinusoidal waves, one with frequency $f_c = n_c/4T_b$ for any fixed integer n_c, and the other with frequency $1/4T_b$. This generates two phase coherent sinusoidal waves at frequency f_1 and f_2, which are related to the carrier frequency f_c and the bit rate $1/T_b$ from (4) and (5) for $h = 1/2$. These two sinusoidal waves are divided by two narrowband filters, one centered at f_1 and the other at f_2. The resulting filter outputs are next linearly combined to produce the pair of quadrature carriers or orthonormal basis functions $\phi_1(t)$ and $\phi_2(t)$. Finally, $\phi_1(t)$ and $\phi_2(t)$ are multiplied with two binary waves $a_1(t)$ and $a_2(t)$, both of which have a bit rate equal to $1/2T_b$.

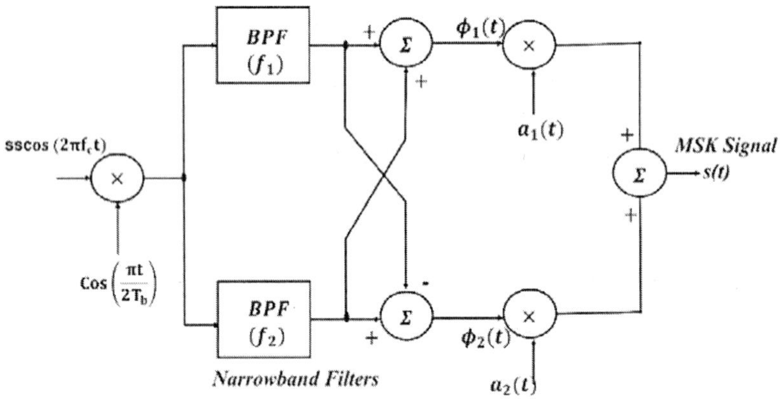

Fig. 5.28 Block diagram of MSK transmitter

Fig. (5.29) shows the block diagram of the coherent MSK receiver. The received signal *x(t)* is correlated with $\phi_1(t)$ and $\phi_2(t)$. In both cases, the integration interval is $2T_b$ seconds, and the integration in the quadrature channel is delayed by T_b seconds with respect to that in the in-phase channel. The resulting in-phase and quadrature channel correlator outputs, x_1 and x_2, are each compared with a threshold of zero; estimates of the phase $\theta(0)$ and $\theta(T_b)$ are then derived in the manner described previously. Finally, these phase decisions are interleaved to estimate the original binary sequence at the transmitter input with the minimum average probability of symbol error in an AWGN channel.

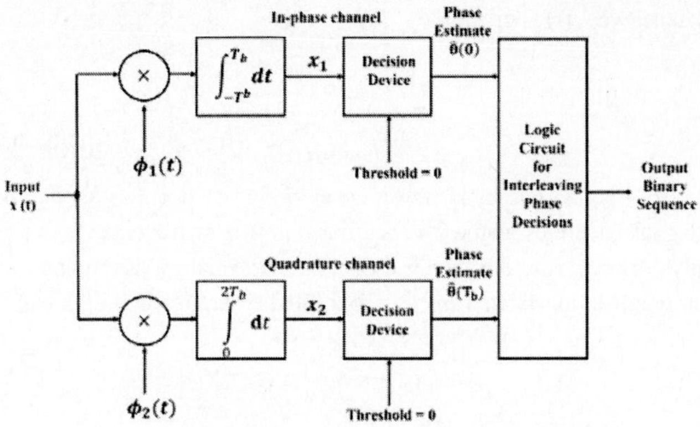

Fig. 5.29 Block diagrams of MSK receiver

5.5.3 Power Spectra of MSK Signals

We assume that the input binary wave is random, with symbols 1 and 0 being equally likely to be delivered, and that the symbols sent during consecutive time slots are statistically independent, much like with the binary FSK signal. These presumptions lead us to three conclusions:

1. The in-phase component equals +*g*(*t*) or –*g*(*t*), depending on the value of phase state θ(0), where the pulse-shaping function

$$\phi(t) = \begin{cases} \sqrt{\frac{2E_b}{T_b}} cos\left(\frac{\pi t}{2T_b}\right), & -T_b \le t \le T_b \\ 0, & otherwise \end{cases} \tag{5.24}$$

g(*t*) has energy spectral density of

$$\psi_g(f) = \frac{32E_bT_b}{\pi^2}\left[\frac{cos(2\pi T_b f)}{16T_b^2 f^2 - 1}\right] \tag{5.25}$$

The power spectral density of the in-phase component equals $\psi_g(f)/2T_b$.

2. The quadrature component equals +*g*(*t*) or –*g*(*t*), depending on the value of the phase state $\theta(T_b)$, where we now get

$$\phi(t) = \begin{cases} \sqrt{\frac{2E_b}{T_b}} sin\left(\frac{\pi t}{2T_b}\right), & 0 \le t \le 2T_b \\ 0, & otherwise \end{cases} \tag{5.26}$$

Despite the different ways that (5.24) and (5.26) define the time gap between two consecutive time slots, we still obtain the same energy spectral density (5.27). As a result, the power spectral density of the in-phase and quadrature components is the same.

3. Since the MSK signal's in-phase and quadrature components are statistically independent, the baseband power spectral density of s(t) is given by

$$S_B(f) = 2\left(\frac{\psi_g(f)}{2T_b}\right) \tag{5.27}$$

$$= \sqrt{E_b}\cos[\theta(0)], \qquad -T_b \leq t \leq T_b$$

For $f >> 1/T$b, The QPSK signal baseband power spectral density declines as the inverse square of frequency, whereas the MSK signal's declines as the inverse fourth power of frequency. As a result, MSK produces less interference outside the signal band of relevance than QPSK. This is a desirable property of MSK, particularly when the digital communication system runs in a congested environment with a bandwidth restriction.

References and Suggested Reading

1. B. P. Lathi, and Zhi Ding, "Modern Digital and Analog Communication Systems", Oxford University Press, 2010.
2. Simon Haykin, "Communication Systems", Wiley, 2007.
3. Herbert Taub, Donald L. Schilling, and Gautam Saha, "Principal of Communication Systems", McGraw Hill Education, 2017.
4. P.F. Panter, "Modulation, Noise and Spectral Analysis", McGraw-Hill, New York, 1965.
5. https://nptel.ac.in/courses/117102059
6. https://ocw.mit.edu/courses/16-36-communication-systems-engineering-spring-2009/pages/lecture-notes/
7. George Clifford Hartley, "Techniques of Pulse-code Modulation in Communication Networks", 19.

6 Passband Receiver

UNIT SPECIFICS

Through this unit we have discussed the following aspects:

- *Review of digital modulation tradeoffs,*
- *Optimum demodulation of digital signals over band-limited channels*
- *Maximum likelihood sequence detection (vertibi receiver).*
- *Equalization Techniques,*
- *Synchronization and carrier recovery for digital modulation,*

The practical applications of the topics are discussed for generating further curiosity and creativity as well as improving problem solving capacity.

Besides giving a large number of multiple-choice questions as well as questions of short and long answer types marked in two categories following lower and higher order of Bloom's taxonomy, assignments through a number of numerical problems, a list of references and suggested readings are given in the unit so that one can go through them for practice. It is important to note that for getting more information on various topics of interest some QR codes have been provided in different sections which can be scanned for relevant supportive knowledge.

After the related practical, based on the content, there is a "Know More" section. This section has been carefully designed so that the supplementary information provided in this part becomes beneficial for the users of the book. This section mainly highlights the initial activity, examples of some interesting facts, analogy, history of the development of the subject focusing the salient observations and finding, timelines starting from the development of the concerned topics up to the recent time, applications of the subject matter for our day-to-day real life or/and industrial applications on variety of aspects, case study related to environmental, sustainability, social and ethical issues whichever applicable, and finally inquisitiveness and curiosity topics of the unit.

RATIONALE

This fundamental unit on analog modulation techniques helps students to get a primary idea about the different types of amplitude and angle modulation techniques and their application in communication process. We have started with basics and pre-requisites required for the topic. First, signals and systems is thoroughly revised before moving on to the more complex topics. Then we have covered the amplitude modulation, its types and demodulation process. Similarly, then angle modulation and demodulation is covered. In angle modulation phase and frequency modulation are thoroughly explained. The chapter is concluded with some MCQs and numerical questions.

Modulation is the most important step in transmitting an information bearing signal over a bandpass channel, such as a telephone line, satellite channel etc.. It accomplishes the task of shifting the range of frequencies of the signal to some another frequency range suitable for transmission. We will learn more about the modulation process in this chapter.

PRE-REQUISITES

Mathematics: Calculus (Class XII)

Physics: Basic of Communication (Class XII)

UNIT OUTCOMES

List of outcomes of this unit is as follows:

U1-O1: Describe basics digital modulation tradeoffs.

U1-O2: Explain the optimum demodulation of digital signals over band-limited channels.

U1-O3: Describe maximum likelihood sequence detection (vertibi receiver).

U1-O4: Explain equalization techniques.

U1-O5: Synchronization and carrier recovery for digital modulation.

This table below need to be filled after discussion as marked with red.

Unit-1 Outcomes	EXPECTED MAPPING WITH COURSE OUTCOMES (1- Weak Correlation; 2- Medium correlation; 3- Strong Correlation)					
	CO-1	CO-2	CO-3	CO-4	CO-5	CO-6
U1-O1	3	3	3	-	3	1
U1-O2	1	1	2	2	1	-
U1-O3	2	1	3	1	2	1
U1-O4	-	-	3	1	2	2
U1-O5	3	3	3	-	3	1

6.1 DIGITAL MODULATION TRADE-OFFs

As we learned in the previous chapters, we must lessen quantization distortion and violations caused by channel noise for a tolerable signal quality at the receiver end of the designed digital communication system. The cost of the system must also be considered. For this the following trade-off are to be considered.

Bit Rate or Bandwidth (BW): To transmit high bit rates more BW is required but they also reduce distortion due to quantization. With the growth of mobile data traffic driven by smart phones and tablet PCs there is bandwidth challenge which needs to be addressed for the upcoming 5G and beyond technology. Fig. 6.1 shows forecast of the number of internet users in India from 2010 to 2025 estimated by Statista's Key Market Indicators [1].

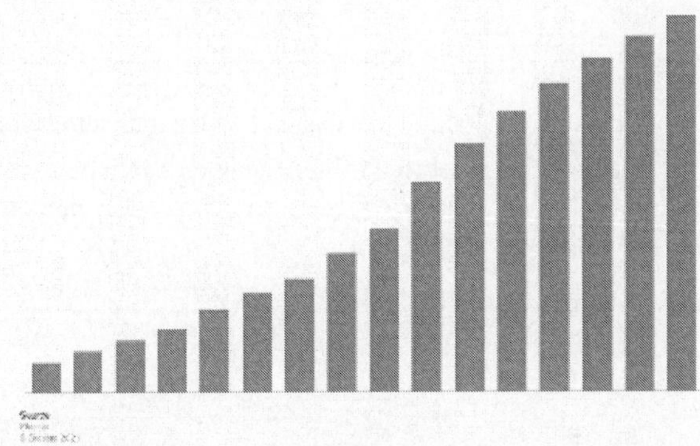

Fig. 6.1: Forecast of the number of internet users in India from 2010 to 2025 [1]

Signal Energy/Bit: Bit width or signal energy per bit are reduced with higher bit rate. As a result, there are more bit errors or violations on transmission channels, which lowers the signal quality at the receiver end as channel noise-related distortion increases. Therefore, while choosing bit rate, we must make a trade-off between acceptable quantization distortion and acceptable channel noise distortion. The amount of spectrum used, however, restricts the number of users. The same information can also be transmitted over a smaller bandwidth by using more complicated transmitters and receivers. This fundamental trade-off is represented in fig. 6.2.

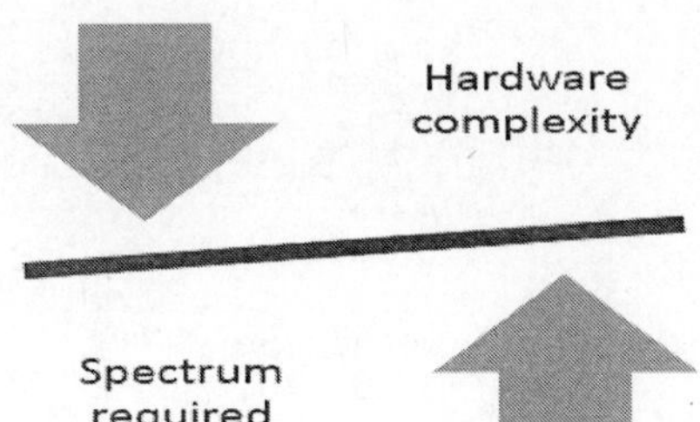

Fig. 6.2 : The fundamental trade-off

Circuit complexity: As we have seen in the earlier chapters, complex circuit of DPCM and ADPCM reduces quantization distortion even at less bit rates hence less distortion due to channel noise but cost of user equipment in case of DPCM and ADPCM is very high as compared to PCM.

6.2 OPTIMUM DEMODULATION OF DIGITAL SIGNALS OVER BAND-LIMITED CHANNELSS

The pulse shapes that will result in zero inter symbol interference (ISI) at the receiver, allow us to achieve maximum possible transmission rate with zero ISI.

The Condition for zero ISI

$$x(t = kT) = x_k = 1,\ k = 0 \tag{6.1}$$

$$x_k = 0,\ k \neq 0$$

Consider transmitting $x(t)$ data at rate $\frac{1}{T}$ through a channel with bandwidth W or through a distorting channel. Our aim is to find the optimum receiver. A filter acting on a infinite sequence of impulse with impulse response $f(t)$ is assumed. The channel is characterized by an impulse response of g(t) and the receiver is a filter sampled at rate 1/T with impulse response h(t).

The transmitted signal is given by

$$S(t) = \sum_{m=-N}^{N} u_m\, f(t - mT) \tag{6.2}$$

where u_m is the data symbol transmitted during the mth signaling interval assumed to be in the alphabet A and f(t) is the waveform used for transmission.

The output of channel filter

$$z(t) = \oint_{\infty}^{-\infty} g(t - \tau)s(\tau)d\tau \tag{6.3}$$

Substituting (6.2) in (6.3) and after some simplification we get

$$z(t) = \sum_{m=-N}^{N} u_m\, h(t - mT) \tag{6.4}$$

where $h(t) = \int_{-\infty}^{\infty} g(t - \tau)f(\tau)d\tau$.

Th received term consists of two terms, i.e., signal and noise

$$r(t) = \sum_{m=-N}^{N} u_m\, h(t - mT) + n(t) \tag{6.5}$$

Since *n(t)* is white gaussian noise the optimum detector to compute each data set *v*

$$\Lambda_N = 2\left(r(t), S_v(t) - ||S_v||^2\right) \tag{6.6}$$

$$\Lambda_N = 2\oint_{-\infty}^{\infty} r(t)S_v(t)dt - \int_{-\infty}^{\infty} S_v^2(t)dt$$

$$\Lambda_N = 2\int_{-\infty}^{\infty} r(t) \sum_{m=-N}^{N} v_k\, h(t-kT)dt - \int_{-\infty}^{\infty} \sum_{m_{k_k}^{\Sigma v}} v_m\, h(t-kT)h(t-mT)dt$$

After simplification

$$\Lambda_N = 2\sum_{k=-N}^{N} v_k\, y_k - \sum_{k=-N}^{N}\sum_{m=-N}^{N} v_k\, v_m x_{m-k} \tag{6.7}$$

Where $y_k = \int_{-\infty}^{\infty} r(t)h(t-kT)dt$ and $x_{m-k} = \int_{-\infty}^{\infty} h(t-mT)h(t-kT)dt.$

Thus, the optimum decision rule is

$$chose\ v\ if\ \Lambda_N(v) = \max_{u} \Lambda_N(u) \tag{6.8}$$

6.2.1 Optimum Demodulation of Digital Signals Over Band- Limited-Maximum Likelihood Sequence Detection (Viterbi Receiver)

The symbol-by-symbol detector is employed when the signal has no memory to minimize the symbol error probability. The optimal detector, however, is one that makes decisions based on observations of a sequence of received signals over subsequent signal intervals when the transmitted signal has memory. The maximum likelihood sequence detection algorithm is described in this section, using the trellis which minimize search the minimum Euclidean distance.

Maximum likelihood sequence detection (Viterbi receiver): Viterbi in 1967 proposed as Maximum likelihood sequence detection (MLSD) algorithm for digital communication system. The MLSD algorithm computation and memory requirement grow linearly with respect to the message length *K*. The MLSD algorithm selects the most likely symbol sequence. The issue is reduced to the optimal detection problem if transmission is over K symbol intervals, and each path through the trellis is K symbols length.

Let the received signal is represented by *r(t)* over the *k* signalling interval. Selecting a K-signal path through the trellis such that the Euclidean distance between the path and r(t) is as minimal as possible corresponds to ML detection.

Since,

$$\int_{0}^{KT} |r(x)-s(x)|^2 dx = \sum_{k=1}^{K}\int_{(k-1)T}^{kT} |r(x)-s(x)|^2 \tag{6.9}$$

The optimal detection is given as:

$$\left(\hat{s}^{(1)}, \hat{s}^{(2)}, \ldots, \hat{s}^{(K)}\right) = \arg \min_{\left(\hat{s}^{(1)}, \hat{s}^{(2)}, \ldots, \hat{s}^{(K)}\right)\in\Gamma} \sum_{k=1}^{K} ||r^k - s^{(k)}|| \tag{6.10}$$

$$= \arg \min_{(\hat{s}^{(1)},\hat{s}^{(2)},\ldots,\hat{s}^{(K)})\in\Gamma} \sum_{k=1}^{K} D(r^k, s^k) \tag{6.11}$$

Where Γ denotes the trellis

Let us consider an example of MLSD, the non-return to zero inverted (NRZI) signaling is considered. The trellis diagram of NRZI is shown in Fig. 6.3. The binary PAM is utilized for each signal transmission. Hence the possible signal transmitted is given as

$$s_1 = -s_2 = \sqrt{\zeta_b} \tag{6.12}$$

Where ζ_b is the energy per bit.

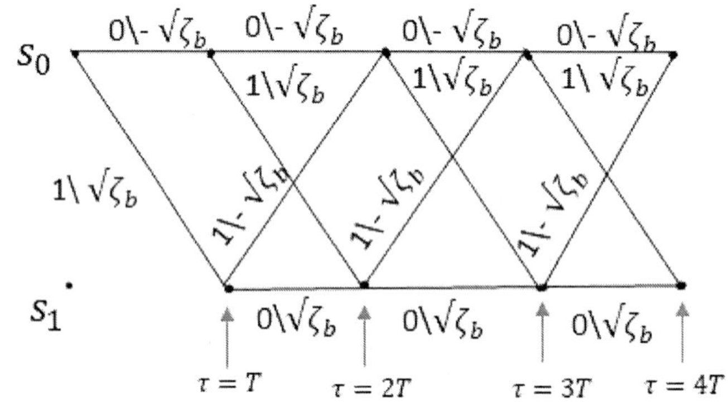

Fig 6.3: NRZI signal Trellis diagram

The Euclidean distance is computed for every possible sequence in search of most likely sequence. Viterbi algorithm is used to eliminate the sequence as new data received from the demodulator. The Viterbi algorithm is a used in convolution decoding, however, we describe here in the context of NRZI detection. The initial state is assumed as S_0. The corresponding trellis is show in Fig 6.3.

The received signal is at time $\tau = T$ and $\tau = T$ is given as $x_1 = s_1^{(m)} + n$ and $x_2 = s_2^{(m)} + n$, respectively. Let $L = 1$ denotes the signal memory of 1 bit, it is observed that trellis reaches its steady state in two transitions. The two paths entering each S_0 at $\tau = 2T$ corresponds to the information bits represented as $\left(-\sqrt{\zeta_b}, -\sqrt{\zeta_b}\right)$ and $\left(\sqrt{\zeta_b}, -\sqrt{\zeta_b}\right)$ for (0, 0) and (1, 1), respectively. The two paths entering each S_1 at $\tau = 2T$ corresponds to the information bits represented as $\left(-\sqrt{\zeta_b}, \sqrt{\zeta_b}\right)$ and $\left(\sqrt{\zeta_b}, \sqrt{\zeta_b}\right)$ for (0, 1) and (1, 0), respectively. The Euclidian distance for the entering node S_0 is given by

$$D_0(0,0) = \left(x_1 + \sqrt{\zeta_b}\right)^2 + \left(x_1 + \sqrt{\zeta_b}\right)^2 \tag{6.13}$$

$$D_0(1,1) = \left(x_1 - \sqrt{\zeta_b}\right)^2 + \left(x_1 + \sqrt{\zeta_b}\right)^2$$

The greater distance is discarded by comparing the two different paths in the matric. Similarly, the Euclidian distance is calculated for the paths entering the S_1

$$D_1(0,1) = \left(x_1 + \sqrt{\zeta_b}\right)^2 + \left(x_1 - \sqrt{\zeta_b}\right)^2 \tag{6.14}$$

$$D_1(1,0) = \left(x_1 - \sqrt{\zeta_b}\right)^2 + \left(x_1 - \sqrt{\zeta_b}\right)^2$$

The two metrics are compared, and the signal path with the larger metric is eliminated. Thus, at $\tau = T$, we are left with two survivor paths, one at node So and the other at node S1 and their corresponding metrics.

At $\tau = 3T$, x_3 is received. The metrics of two path arriving the S_0 is calculated. The survivors at $\tau = 2T$ are (0, 0) and (0, 1) at S_0 and S_1 respectively. The at $\tau = 3T$, the two paths entering at S_0 is given as

$$D_0(0,0,0) = D_0(0,0) + \left(x_3 + \sqrt{\zeta_b}\right)^2 \tag{6.15}$$

$$D_0(0,1,1) = D_1(0,1) + \left(x_3 + \sqrt{\zeta_b}\right)^2$$

After comparing the path with larger distance is eliminated. At $\tau = 3T$ the paths entering S_1 is given as

$$D_0(0,0,1) = D_0(0,0) + \left(x_3 - \sqrt{\zeta_b}\right)^2 \tag{6.16}$$

$$D_0(0,1,0) = D_1(0,1) + \left(x_3 - \sqrt{\zeta_b}\right)^2$$

Thus, the Viterbi algorithm compute the and eliminate the greater distance path in each node. And the survivor paths are then extended forward to the next node. Therefore, the number of paths searched in the trellis is reduced by a factor of two at each stage.

6.2 EQUALIZATION TECHNIQUES

Equalization techniques are utilized within a receiver to eliminate the effects of inter symbol interference (ISI) which arises mainly because of dispersive nature of the channel resulting in spreading of the transmitted pulse and hence overlapping of the adjacent pulses. ISI creates a major problem to achieve high speed transmissions over radio channels. Equalizers mainly attempts to reverse this distortion created and recover the transmitted symbol. They should adapt to the channel conditions which are mostly time varying and unknown.

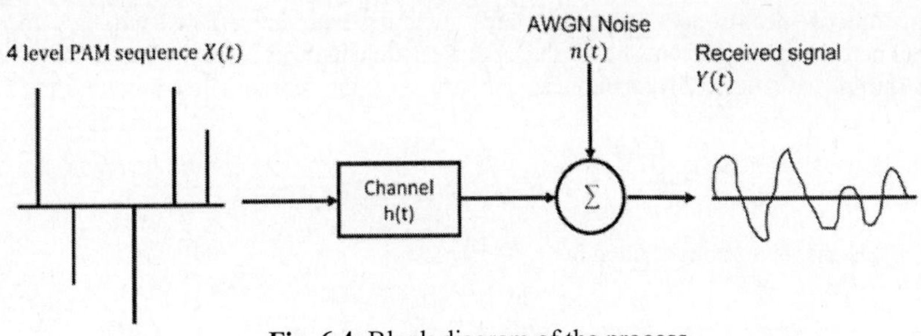

Fig. 6.4: Block diagram of the process

Figure 6.4 depicts a message signal $X(t)$ with four level pulse amplitude modulation (PAM) which is transmitted through a communication channel having impulse response equal to $h(t)$. The signal is also affected by AWGN noise $n(t)$ at the receiver end. Clearly, the received signal $Y(t)$ can be observed to be a distorted signal in comparison to the transmitted message pulse.

Equalization techniques can be divided into two main categories namely maximum likelihood sequence estimation (MLSE) and equalization with filters. The MLSE method (for example Viterbi equalization) performs the measurement of impulse response of the channel and then adjusts the receiver according to the same. On the other hand, equalization with filters utilize either a simple linear filter or some complex algorithm for the distorted pulses. Some of the different types of such equalizers are as follows:

a. **Linear Equalizer**: A linear filter is used to process the signal. Designed with finite impulse response filter, they are also called as linear transversal filters.

b. **Zero Forcing (ZF) Equalizer**: It is a form of linear equalizer which uses the inverse of channel's frequency response for the received signal. The idea is to bring down the ISI to zero level in a noise-free scenario or in cases where the ISI is more significant than the noise.

c. **Decision Feedback Equalizer**: It is a non-linear equalizer which corrects the present symbol based on estimates made about the previous symbols (i.e. high/low). DFE has a main advantage in comparison to the linear equalizers in that it cancels ISI without any amplification of the noise.

d. **Blind Equalizer**: This technique estimates the transmitted signal utilizing only the knowledge of the transmitted signal statistics.

e. **Viterbi Equalizer**: With the aim of minimizing the probability of error during the entire sequence, it determines the equalization problem's optimal solution.

f. **BCJR Equalizer**: It makes use of the Bahl–Cocke–Jelinek–Raviv (BCJR) algorithm the objective of which is to reduce the probability of incorrect estimation of a given bit.

6.3.1 Optimum Linear Receiver

In a practical communication scenario, both channel noise (that leads to design of matched filter) and ISI (that leads to design of pulse shaping transmit filter to realize Nyquist channel) together affect the signal transmission. Thus, an optimum receiver needs to be designed considering that the linear channel is both noisy and dispersive. Equalizers like ZF are though easy to design but the effect of noise is ignored. This leads to noise enhancement and thus overall performance degradation. A better approach for receiver design is the use of mean-square criterion which includes both the effects and thus for a given computational complexity it performs better than ZF. Such an equalizer is referred to as minimum-mean square error (mmse) equalizer. Given impulse response $c(t)$, the receive filter produces the following response

$$y(t) = \int_{-\infty}^{\infty} c(\tau)\, x(t-\tau)\, d\tau \tag{6.17}$$

where x(t) is channel output defined by

$$x(t) = \sum_{k} a_k\, q(t - k\,T_b) + w(t) \tag{6.18}$$

where a_k is transmitted symbol at $t = k\,T_b$ time, T_b is bit duration, and $w(t)$ represents the channel noise. The function $q(t)$ is convolution of $g(t)$ and $h(t)$ which are impulse responses of pulse-shaping filter and channel, respectively. Utilizing the above two equations (6.17) and (6.18) and sampling $y(t)$ at time $t = i\,T_b$, we obtain

$$y(i\,T_b) = \varepsilon_i + n_i \tag{6.19}$$

where signal component ε_i and noise component n_i are defined by

$$\varepsilon_i = \sum_k a_k \int_{-\infty}^{\infty} c(\tau)\, q(i\,T_b - k\,T_b - \tau)\, d\tau \tag{6.20}$$

$$n_i = \int_{-\infty}^{\infty} c(\tau)\, w(i\,T_b - \tau)\, d\tau \tag{6.21}$$

For perfect operation of receiver, $y(i\,T_b)$ should be equal to transmitted symbol a_i. Thus, the error signal is defined as

$$\begin{aligned} e_i &= y(i\,T_b) - a_i \\ &= \varepsilon_i + n_i - a_i \end{aligned} \tag{6.22}$$

Accordingly, mean-square error can be defined as

$$J = \frac{1}{2} E[e_i^{\,2}]$$

$$J = \frac{1}{2} E[\varepsilon_i^{\,2}] + \frac{1}{2} E[n_i^{\,2}] + \frac{1}{2} E[a_i^{\,2}] + E[\varepsilon_i\, n_i] - E[a_i\, n_i] - E[\varepsilon_i\, a_i] \tag{6.23}$$

where $E[.]$ is statistical expectation operator.

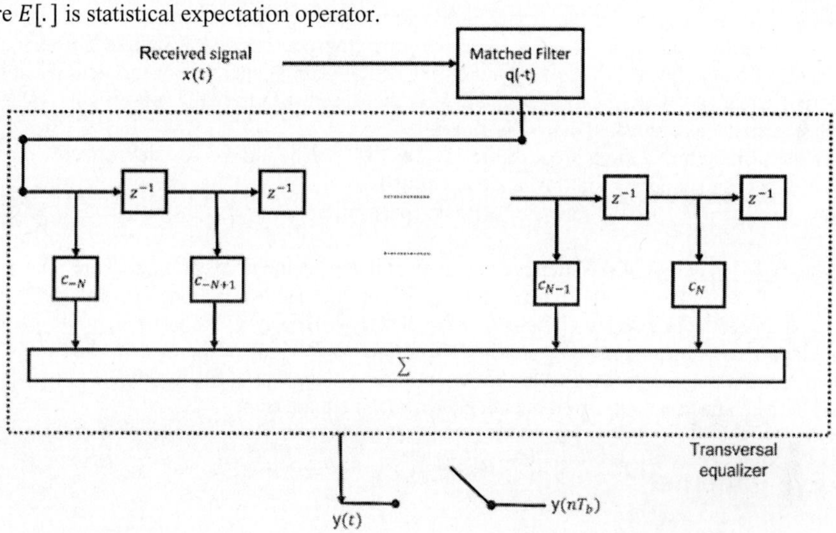

Fig. 6.5 A MMSE receiver with cascaded transversal equalizer and matched filter.

The MMSE receiver shown in Fig. 6.2 consisting of cascaded transversal equalizer and matched filter works well when we have access to the system to be equalized such that a transversal equalizer characterized by set of coefficients $\{c_k\}_{k=-N}^{N}$. However, the channel is mostly time-varying and designing matched filter and equalizer based on average characteristics of channel may not be sufficient to reduce the detrimental effects of ISI and noise. This leads to the need of designing equalizers where coefficients automatically adjusted according to a built-in algorithm.

6.3.2 Adaptive Equalizer

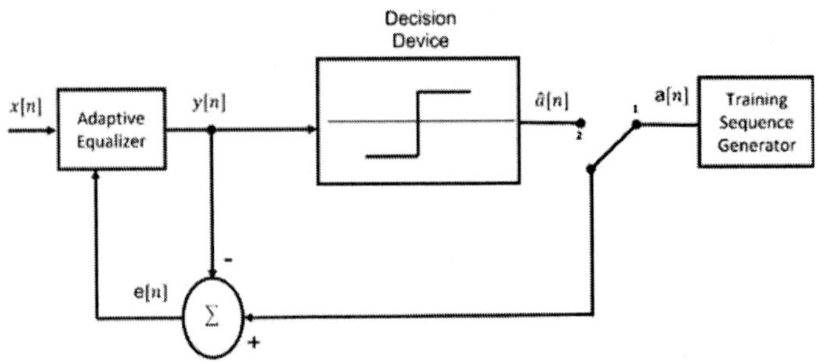

Fig. 6.6 Training and tracking mode of adaptive equalizer.

The equalization techniques can be classified either as preset or adaptive. The assumption made for preset equalizers is that the communication channel is time invariant. On the other hand, adaptive equalizers consider the channel to be slowly varying with time and therefore equalizer filters are designed such that their coefficients are varying in accordance with changes in the channel.

Adaptive equalizers work in two different modes namely the decision-directed mode and the training mode, as shown Fig 6.3, where the switch is moved on position 1 for training mode and is moved to position 2 for tracking mode. In training mode, a known PN sequence is transmitted from the transmitter and the same is generated at the receiver after a time shift equal to delay in transmission which is given to the adaptive equalizer as the actual response. It is then used to adjust the tap-weights of equalizer in accordance with the least-mean-square algorithm.

After finishing the training process, the equalizer is switched to the decision-directed mode of operation. The error signal $e[n]$ in this mode is defined as

$$e[n] = \hat{a}[n] - y[n] \tag{6.24}$$

where $y[n]$ is output of the adaptive equalizer and $\hat{a}[n]$ is the final estimate of a$[n]$ (i.e. transmitted symbol). It is to note that the decision-directed mode can track slow variations in the channel. Also, if the step-size parameter is larger, the equalizer can track fastly. However, the large step-size parameter can lead to a high excess mean-square error which is defined as the part of the mean square error exceeding minimum value when the tap weights are set to optimum value. Therefore, the choice of the step-size is a trade-off between fast tracking and reducing excess mean-square error.

6.3.3 Eye Patterns

An eye pattern provides a lot of useful insights on the transmission system performance. It is basically a synchronized superposition of all possible realizations of signal of interest within a signalling interval which resembles the human eye. The width of the eye-opening shows the time interval where sampled received signal is free from ISI. The rate of closure of eye with respect to variations in sampling time shows the sensitivity of system to timing errors. Further, the hight of the eye opening denotes the system noise margin. When the effect of ISI is severe, traces from lower portion cross traces from upper portion of the eye pattern resulting in the eye to be closed.

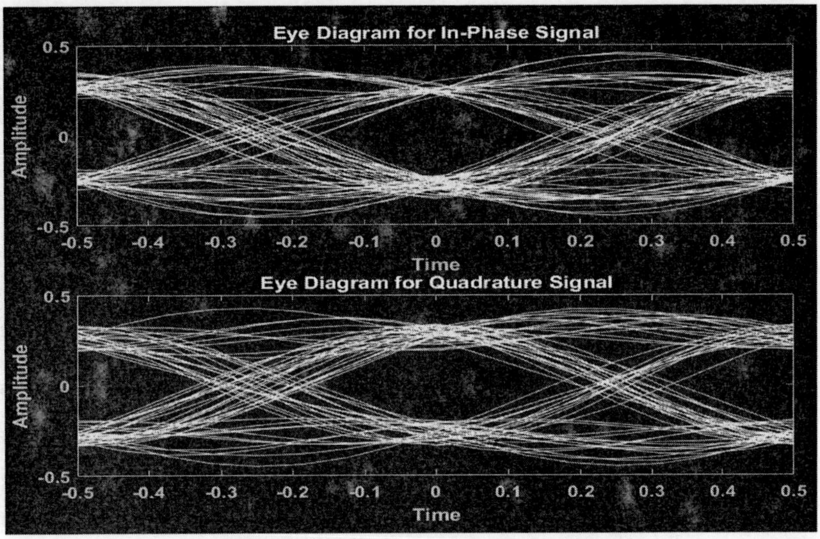

Fig. 6.7: Eye diagram of transmitted signal.

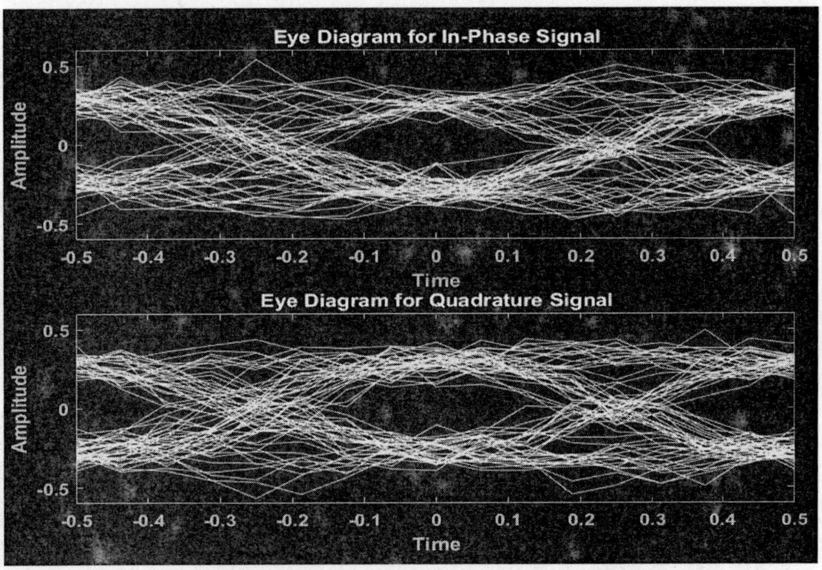

Fig. 6.8: Eye diagram of received signal at SNR 20 dB.

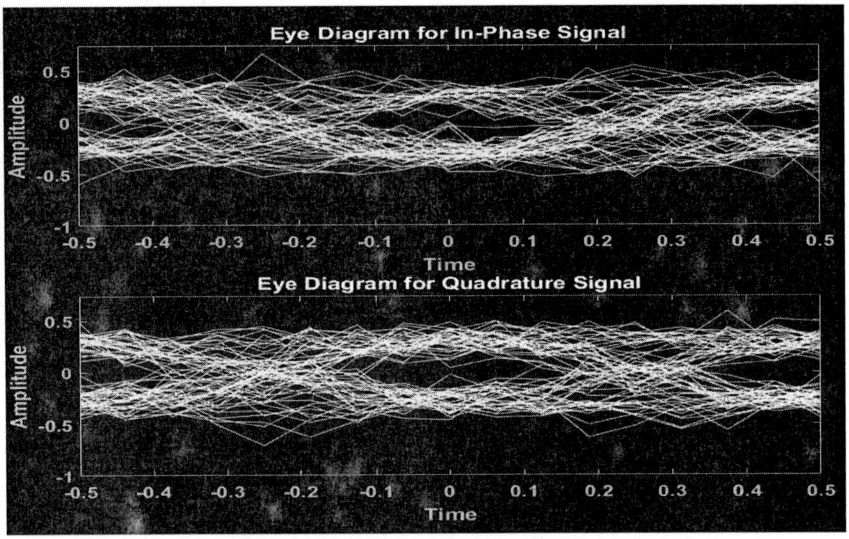

Fig. 6.9: Eye diagram of received signal at 15 dB.

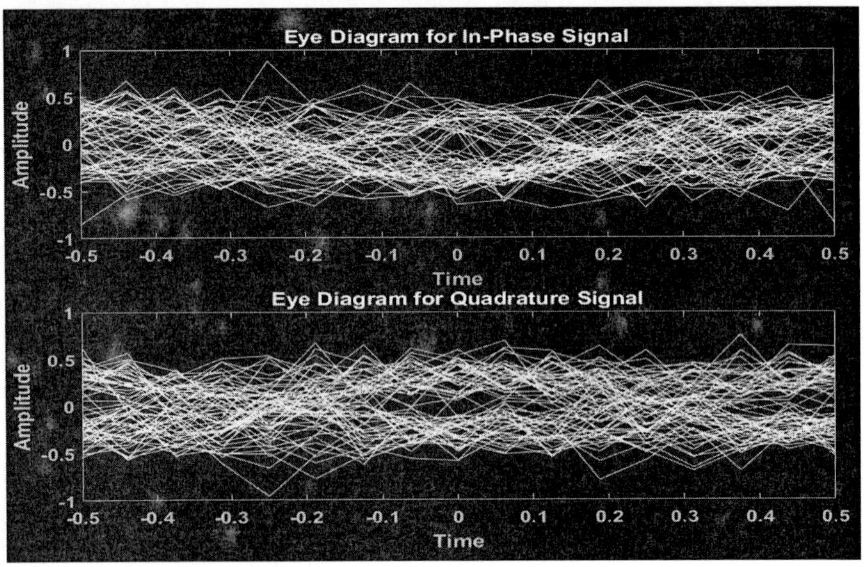

Fig. 6.10: Eye diagram of received signal at 10 dB.

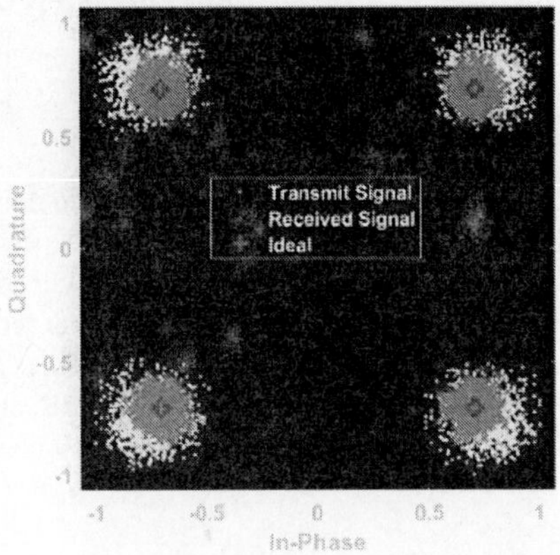

(a)

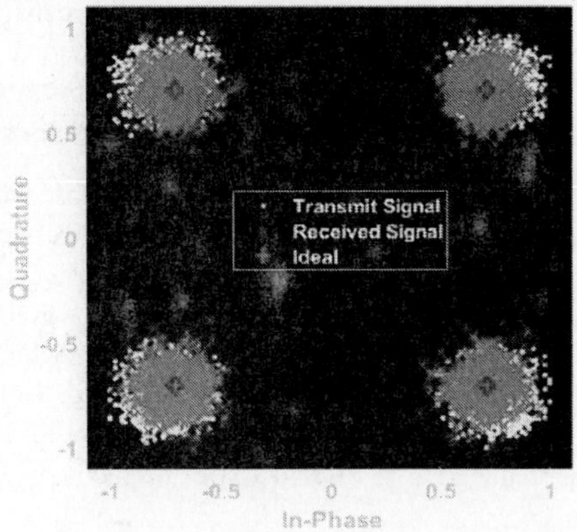

(b)

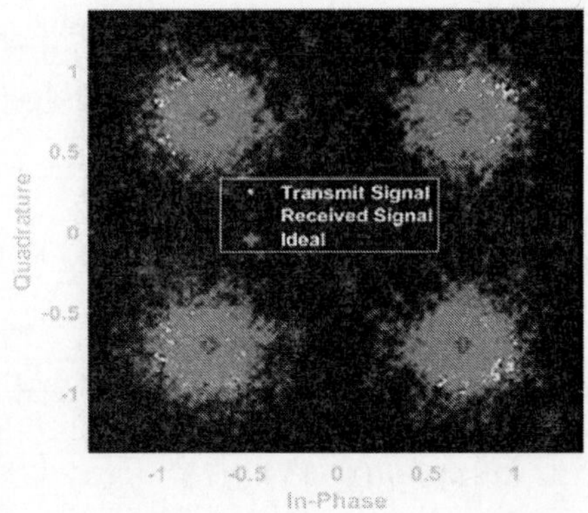

(c)

Fig. 6.11 Scatter plot of transmitted and received signal at SNR (a) 20 dB, (b) 15 dB and (c) 10 dB.

Fig. 6.7, 6.8, 6.9 and 6.10 show the eye diagram of transmitted, received signal at SNR 20 dB, 15 dB and 10 dB, respectively. An ideal eye diagram shows a wider eye having a lot of margins in both horizontal and vertical direction as shown in Fig. 6.7 which infers lowest possible error rate in the receiver decisions. When the QPSK symbols are transmitted through an AWGN channel, some distortion is expected which we can observe in Fig. 6.8, 6.9 and 6.10.

Eye diagram in Fig. 6.8 corresponds to the case when SNR is 20 dB and thus the distortion due to noise is evident in comparison to Fig. 6.7. The eye is less wide which will impact the bit error rate (BER) performance. On further reducing the SNR to 15 dB, the eye diagram of received signal is shown by Fig. 6.9. On comparing the same with Fig. 6.8, the width of the eye is gradually reduced. Further, when the SNR is further reduced to 10 dB, the eye is observed to be almost close which depicts the case of low BER performance. This can also be verified from the scatter plot in Fig. 6.11. Fig. 6.11 (a) corresponds to the case when SNR is 20 dB, and it can be observed that the symbol mapping is more localized compared to that in Fig. 6.11 (b) and (c) when SNR is further reduced to 15 dB and 10 dB, respectively.

6.4 SYNCHRONIZATION AND CARRIER RECOVERY OF DIGITAL MODULATION

To recover the sent data in a digital communication system, the output of the demodulator must be sampled repeatedly, once at every symbol interval. Generally, there is a propagation delay between the transmitter and the receiver, and that is unknown to the receiver. To synchronize the sample of the demodulator's output, symbol timing must be determined from the received signal. If the detector is phase-coherent, the carrier offset caused by the propagation delay in the transmitted signal must be approximated at the receiver. The digital communication system that transfers information simultaneously needs symbol synchronization. If the signal is detected coherently, then carrier recovery is necessary.

6.4.1 Estimation of Signal Parameter

Let the signal is transmitted through the channel that introduces delay and adds the Gaussian noise. Hence, the received signal can be expressed as

$$y(t) = m(t-\tau) + n(t) \tag{6.25}$$

where *m(t)* denotes the bandpass signal, τ denotes the processing delay, and *n(t)* denotes the additive white Gaussian noise. Further, the received signal can be expressed as

$$y(t) = \mathrm{Re}\left\{\left[m_l(t-\tau)e^{j\phi} + z(t)\right]e^{j2\pi f_c t}\right\} \tag{6.26}$$

where $m_l(t)$ denotes the equivalent low-pass signal, $\phi = -2\pi f_c t$ denotes the carrier phase, and f_c denotes the carrier frequency. Since the carrier phase can be estimated from the knowledge of the carrier phase and propagation delay. Hence, the propagation delay needs to be estimated. However, the carrier phase does not depend only on the time delay because the oscillator at the receiver that produces the carrier signal for demodulation is typically not synchronous in phase with the oscillator at the transmitter. The two oscillators might steadily drift over time. The symbol interval *T* determines how precisely one must synchronize in time to demodulate the received signal. The estimation error in estimating the processing delay τ typically has to be a little percentage of *T*. For practical purposes, 1% of *T* is sufficient. However, because f_c is typically large, this level of precision is typically insufficient for determining the carrier phase. To demodulate and coherently detect the received signal, we need to estimate both parameters τ and ϕ. Hence, the received signal can be expressed as:

$$y(t) = m(t;\ \phi,\ \tau) + n(t) \tag{6.27}$$

where τ and ϕ are the parameters to be estimated. The maximum likelihood (ML) and maximum posteriori probability (MAP) criteria are used for the signal parameter estimation. For simplicity, let € denote the parameter vector τ and ϕ. The a priori probability density function p(€) is used in the MAP criteria models because the € is considered random whereas € is as deterministic but unknown in the ML criteria. In our case, the parameters τ and ϕ are unknown and deterministic, therefore, we can adopt ML criteria for the estimation. We require the receiver to extract the estimate of the signal parameters by tracking the received signal across the observation interval, which is $T_0 \geq T$. In practice, tracking loops (either analog or digital) are used to continuously update the estimations during the estimation process. To maximize the p(y|€) with respect to the signal parameter τ and ϕ is equivalent to maximizing the likelihood function $\Lambda(€)$ which can be given as

$$\Lambda(€) = \exp\left\{ -\frac{1}{N_0} \int_{T_0} \left[y(t) - m(t;€) \right]^2 dt \right\} \tag{6.28}$$

Now, we consider the signal parameter estimation to maximize the $\Lambda(€)$.

6.4.2 Carrier Recovery and Symbol Synchronization in Signal Demodulation:

For binary phase shift keying demodulator and detector, the correlator's reference signal is produced by using the carrier phase estimation. The sampler and the output of the signal pulse generator are both under the control of the symbol synchronizer. The signal generator can be removed if the signal pulse is rectangular.

There are two fundamental methods for handling carrier synchronization at the receiver:

One method is multiplex, usually in frequency, a special signal known as a pilot signal, which enables the receiver to extract and synchronize its local oscillator to the carrier frequency and phase of the received signal. The receiver uses a phase-locked loop (PLL) to track and acquire an unmodulated carrier component that is sent along with the information-bearing signal. The PLL is made to have a narrow bandwidth so that the presence of frequency components from the information-bearing signal does not significantly affect it.

Deriving the carrier phase estimate directly from the modulated signal is the second strategy, which appears to be more common in reality. The main benefit of this strategy is that the information-carrying signal will be transmitted using the total transmitter power.

To recover the sent data in a digital communication system, the output of the demodulator must be sampled repeatedly, once at every symbol interval. Generally, there is a propagation delay between the transmitter and the receiver, and that is unknown to the receiver. To synchronize the sample of the demodulator's output, symbol timing must be determined from the received signal. If the detector is phase-coherent, the carrier offset caused by the propagation delay in the transmitted signal must be approximated at the receiver. The digital communication system that transfers information simultaneously needs symbol synchronization. If the signal is detected coherently, then carrier recovery is necessary.

6.4.3 Recursive Algorithm For Maximum Likelihood Estimation of the Carrier Phase

To solve the synchronization problem, first of all, we have to formulate the log-likelihood function of the carrier phase Θ. By utilizing the adaptive filtering approach, the updated estimate can be formulated as

$$\hat{\theta}[n+1] = \hat{\theta}[n] + Ye[n] \tag{6.29}$$

where $\hat{\theta}[n+1]$ denotes the updated estimate of the carrier phase θ, $\hat{\theta}[n]$ denotes the old estimate of θ, Y denotes the step-size parameter, and e[n] denotes the error signal. Fig. 1 shows the first-order digital filter. From Fig. 1, it is shown that, at the output of the matched filter, the detector applies an estimate on the transmitted symbol. The error generator generates the error signal e[n]. Further, the look-up table provides the value of $exp(-j\hat{\theta}[n])$ for an input $\hat{\theta}[n]$. The unit-delay element provides the delay equal to the symbol period T. Further improvement in the tracking performance of the synchronization system can be designed by cascading two first-order digital filters.

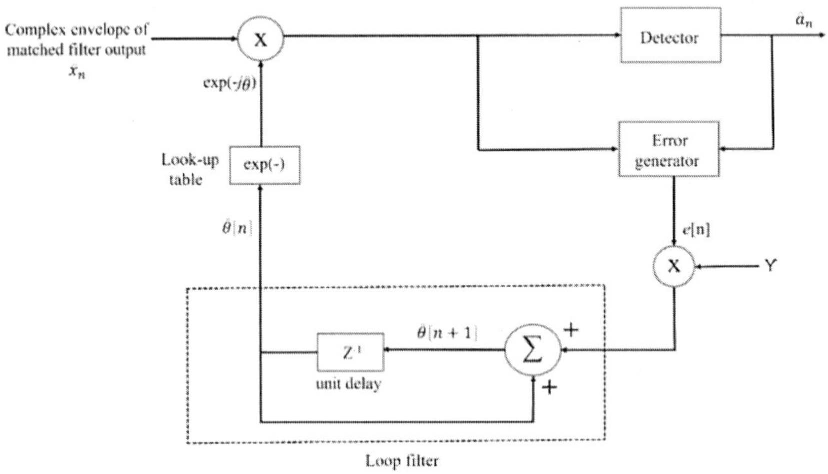

Fig. 6.12: Recursive Costas loop of the first-order filter

References and Suggested Reading

1. B. P. Lathi, and Zhi Ding, "Modern Digital and Analog Communication Systems", Oxford University Press, 2010.

2. Simon Haykin, "Communication Systems", Wiley, 2007.

3. Herbert Taub, Donald L. Schilling, and Gautam Saha, "Principal of Communication Systems", McGraw Hill Education, 2017.

4. P.F. Panter, "Modulation, Noise and Spectral Analysis", McGraw-Hill, New York, 1965.

5. https://nptel.ac.in/courses/117102059

6. https://ocw.mit.edu/courses/16-36-communication-systems-engineering-spring-2009/pages/lecture-notes/

7. George Clifford Hartley, "Techniques of Pulse-code Modulation in Communication Networks", 1967.

APPENDIX 1

TABLE 1 Summary of properties of the Fourier transform:

Property	Mathematical Description
1. Linearity	$ag_1(t) + bg_2(t) \rightleftharpoons aG_1(f) + bG_2(f)$ where a and b are constants
2. Time scaling	$g(at) \rightleftharpoons \frac{1}{\|a\|} G\left(\frac{f}{a}\right)$
3. Duality	If $g(t) \rightleftharpoons G(f)$ then $G(t) \rightleftharpoons g(-f)$
4. Time shifting	$g(t - t_0) \rightleftharpoons G(f)\exp(-j2\pi f t_0)$
5. Frequency shifting	$\exp(j2\pi f_c t)g(t) \rightleftharpoons G(f - f_c)$
6. Area under g(t)	$\int_{-\infty}^{\infty} g(t)dt = G(0)$
7. Area under G(f)	$g(0) = \int_{-\infty}^{\infty} G(f)df$
8. Differentiation in the time domain	$\frac{d}{dt} g(t) \rightleftharpoons j2\pi f G(f)$
9. Integration in the time domain	$\int_{-\infty}^{t} g(\tau)d\tau \rightleftharpoons \frac{1}{j2\pi f} G(f) + \frac{G(0)}{2} \delta(f)$
10. Conjugate functions	If $g(t) \rightleftharpoons G(f)$ then $g^*(t) \rightleftharpoons G^*(-f)$
11. Multiplication in the time domain	$g_1(t)g_2(t) \rightleftharpoons \int_{-\infty}^{\infty} G_1(\lambda)G_2(f - \lambda)d\lambda$
12. Convolution in the time domain	$\int_{-\infty}^{\infty} g_1(\tau)g_2(t - \tau)d\tau \rightleftharpoons G_1(f)G_2(f)$

TABLE 2 Fourier-transform pairs

Time Function	Fourier Transform
$\text{rect}\left(\frac{t}{T}\right)$	$T\sin c(fT)$
$\text{Sinc}(2Wt)$	$\frac{1}{2W}\text{rect}\left(\frac{f}{2W}\right)$
$\exp(-at)u(t), a > 0$	$\frac{1}{a + j2\pi f}$
$\exp(-a\|t\|), a > 0$	$\frac{2a}{a^2 + (2\pi f)^2}$
$\exp(-\pi t^2)$	$\exp(-\pi f^2)$
$\begin{cases} 1 - \frac{\|t\|}{T}, & \|t\| < T \\ 0, & \|t\| \geq T \end{cases}$	$T\text{sinc}^2(fT)$
$\delta(t)$	1
1	$\delta(f)$
$\delta(t - t_0)$	$\exp(-j2\pi f t_0)$
$\exp(j2\pi f_c t)$	$\delta(f - f_c)$
$\cos(2\pi f_c t)$	$\frac{1}{2}[\delta(f - f_c) + \delta(f + f_c)]$
$\sin(2\pi f_c t)$	$\frac{1}{2j}[\delta(f - f_c) - \delta(f + f_c)]$
$\text{sgn}(t)$	$\frac{1}{j\pi f}$
$\frac{1}{\pi t}$	$-i\text{sgn}(f)$
$u(t)$	$\frac{1}{2}\delta(f) + \frac{1}{i2\pi f}$
$\sum_{i=\infty}^{\infty} \delta(t - iT_0)$	$\frac{1}{T_0}\sum_{n=-\infty}^{\infty} \delta\left(f - \frac{n}{T_0}\right)$

Notes: $u(t)$= unit step function

$\delta(t)$ = delta function or impulse function

rect(t)= rectangular function of unit amplitude and unit duration centered on the origin sgn(t)= signum function

sinc(t)= sinc function

APPENDIX 2

Table 1 Hilbert transform pairs

Time Function	Hilbert Transform
$m(t)\cos(2\pi f_c t)$	$m(t)\sin(2\pi f_c t)$
$m(t)\sin(2\pi f_c t)$	$-m(t)\cos(2\pi f_c t)$
$\cos(2\pi f_c t)$	$\sin(2\pi f_c t)$
$\sin(2\pi f_c t)$	$-\cos(2\pi f_c t)$
$\frac{\sin t}{t}$	$\frac{1-\cos t}{t}$
$\mathrm{rect}(t)$	$-\frac{1}{\pi}\log\left\lvert\frac{t-\frac{1}{2}}{t+\frac{1}{2}}\right\rvert$
$\delta(t)$	$\frac{1}{\pi t}$
$\frac{1}{1+t^2}$	$\frac{t}{1+t^2}$
$\frac{1}{t}$	$-\pi\delta(t)$

APPENDIX 3

Table 1 Trigonometric Identities

$$\exp(\pm j\theta) = \cos\theta \pm j\sin\theta$$

$$\cos\theta = \frac{1}{2}[\exp(j\theta) + \exp(-j\theta)]$$

$$\sin\theta = \frac{1}{2}[\exp(j\theta) - \exp(-j\theta)]$$

$$\sin^2\theta + \cos^2\theta = 1$$

$$\cos^2\theta - \sin^2\theta = \cos(2\theta)$$

$$\cos^2\theta = \frac{1}{2}[1 + \cos(2\theta)]$$

$$\sin^2\theta = \frac{1}{2}[1 - \cos(2\theta)]$$

$$2\sin\theta\cos\theta = \sin(2\theta)$$

$$\sin(\alpha \pm \beta) = \sin\alpha\cos\beta \pm \cos\alpha\sin\beta$$

$$\cos(\alpha \pm \beta) = \cos\alpha\cos\beta \mp \sin\alpha\sin\beta$$

$$\tan(\alpha \pm \beta) = \frac{\tan\alpha \pm \tan\beta}{1 \mp \tan\alpha\tan\beta}$$

$$\sin\alpha\sin\beta = \frac{1}{2}[\cos(\alpha - \beta) - \cos(\alpha + \beta)]$$

$$\cos\alpha\cos\beta = \frac{1}{2}[\cos(\alpha - \beta) + \cos(\alpha + \beta)]$$

$$\sin\alpha\cos\beta = \frac{1}{2}[\sin(\alpha - \beta) + \sin(\alpha + \beta)]$$

Table 2 Series Expansions

Taylor Series $$f(x) = f(a) + \frac{f'(a)}{1!}(x-a) + \frac{f''(a)}{2!}(x-a)^2 + \cdots + \frac{f^{(n)}(a)}{n!}(x-a)^n + \cdots$$
MacLaurin series $$f(x) = f(0) + \frac{f'(0)}{1!}x + \frac{f''(0)}{2!}x^2 + \cdots + \frac{f^{(n)}(0)}{n!}x^n + \cdots$$
Binomial series $$(1+x)^n = 1 + nx + \frac{n(n-1)}{2!}x^2 + \cdots, \lvert nx \rvert < 1$$
Exponential series $$\exp x = 1 + x + \frac{1}{2!}x^2 + \cdots$$
Logarithmic series $$\log(1+x) = x - \frac{1}{2}x^2 + \frac{1}{3}x^3 - \cdots$$
Trigonometric series $$\sin x = x - \frac{1}{3!}x^3 + \frac{1}{5!}x^5 - \cdots$$ $$\cos x = 1 - \frac{1}{2!}x^2 + \frac{1}{4!}x^4 - \cdots$$ $$\tan x = x + \frac{1}{3}x^3 + \frac{2}{15}x^5 + \cdots$$ $$\sin^{-1} x = x + \frac{1}{6}x^3 + \frac{3}{40}x^5 + \cdots$$ $$\tan^{-1} x = x - \frac{1}{3}x^3 + \frac{1}{5}x^5 - \cdots, \lvert x \rvert < 1$$ $$\operatorname{sinc} x = 1 - \frac{1}{3!}(\pi x)^2 + \frac{1}{5!}(\pi x)^4 - \cdots$$

Appendix 4

Table 1 Table of Bessel functions

	$J_n(x)$								
n\x	0.5	1	2	3	4	6	8	10	12
0	0.9385	0.7652	0.2239	-0.2601	-0.3971	0.1506	0.1717	-0.2459	0.0477
1	0.2423	0.4401	0.5767	0.3391	-0.0660	-0.2767	0.2346	0.0435	-0.2234
2	0.0306	0.1149	0.3528	0.4861	0.3641	-0.2429	-0.1130	0.2546	-0.0849
3	0.0026	0.0196	0.1289	0.3091	0.4302	0.1148	-0.2911	0.0584	0.1951
4	0.0002	0.0025	0.0340	0.1320	0.2811	0.3576	-0.1054	-0.2196	0.1825
5	-	0.0002	0.0070	0.0430	0.1321	0.3621	0.1858	-0.2341	-0.0735
6		-	0.0012	0.0114	0.0491	0.2458	0.3376	-0.0145	-0.2437
7			0.0002	0.0025	0.0152	0.1296	0.3206	0.2167	-0.1703
8			-	0.0005	0.0040	0.0565	0.2235	0.3179	0.0451
9				0.0001	0.0009	0.0212	0.1263	0.2919	0.2304
10				-	0.0002	0.0070	0.0608	0.2075	0.3005
11					-	0.0020	0.0256	0.1231	0.2704
12						0.0005	0.0096	0.0634	0.1953
13						0.0001	0.0033	0.0290	0.1201
14						-	0.0010	0.0120	0.0650

Table 2 Error function

u	erf(u)	u	erf(u)
0.00	0.00000	1.10	0.88021
0.05	0.05637	1.15	0.89612
0.10	0.11246	1.20	0.91031
0.15	0.16800	1.25	0.92290
0.20	0.22270	1.30	0.93401
0.25	0.27633	1.35	0.94376
0.30	0.32863	1.40	0.95229
0.35	0.37938	1.45	0.95970
0.40	0.42839	1.50	0.96611
0.45	0.47548	1.55	0.97162
0.50	0.52050	1.60	0.97635
0.55	0.56332	1.65	0.98038
0.60	0.60386	1.70	0.98379
0.65	0.64203	1.75	0.98667
0.70	0.67780	1.80	0.98909
0.75	0.71116	1.85	0.99111
0.80	0.74210	1.90	0.99279
0.85	0.77067	1.95	0.99418
0.90	0.79691	2.00	0.99532
0.95	0.82089	2.50	0.99959
1.00	0.84270	3.00	0.99998
1.05	0.86244	3.30	0.999998

INDEX

Differential detection,129
Decision feedback equalizer,154
Detrimental effects,155
Decision-directed mode,156
Digital modulation trade-offs,149

H

I

L

M

N

O

P

Q

R

S

T

U

V

W

Z